EVOLUTION FOR EVERYONE

EVOLUTION *for* EVERYONE

How Darwin's Theory Can Change
the Way We Think About Our Lives

David Sloan Wilson

DELACORTE PRESS

EVOLUTION FOR EVERYONE
A Delacorte Book / April 2007

Published by
Bantam Dell
A Division of Random House, Inc.
New York, New York

Book design by Glen Edelstein

Material from "Life Expectancy, Economic Inequality, Homicide, and Reproductive Timing in Chicago Neighborhoods," Margo Wilson and Martin Daly, *British Medical Journal*, January 1997, reprinted with permission from BMJ Publishing Group Ltd.

Material from *The Ancestress Hypothesis* by Kathryn Coe is reprinted with permission from Rutgers University Press.

With kind permission of Springer Science and Business Media, material is reprinted from Brown, S. (2000). Evolutionary models of music: from sexual selection to group selection. *Perspectives in Ethology*. F. Tonneau and N. S. Thompson. New York, Kluwer Academic. 13: 231–281.

Orality & Literacy! The technologizing of the word, Walter J. Ong, copyright 1982, Routledge Press.

Excerpts from *World As Laboratory* by Rebecca Lemov. Copyright © 2005 by Rebecca Lemov. Reprinted by permission of Hill and Wang, a division of Farrar, Straus and Giroux, LLC.

Material from *The Universe in a Single Atom: The Convergence of Science and Spirituality* by His Holiness the Dalai Lama. Used by permission of an imprint of The Doubleday Broadway Publishing Group, a division of Random House, Inc.

Library of Congress Cataloging-in-Publication Data
Wilson, David Sloan.
Evolution for everyone: how Darwin's theory can change the way we think about our lives / David Sloan Wilson.
p.   cm.
Includes bibliographical references.
ISBN: 978-0-385-34021-2 (hardcover)
1. Evolution. I. Title.
B818.W665 2007
576.801—dc22   2006023685

Printed in the United States of America
Published simultaneously in Canada

www.bantamdell.com

10 9 8 7 6 5 4 3
BVG

For H.

CONTENTS

EVOLUTION FOR EVERYONE

1

The Future Can Differ from the Past

THIS IS A BOOK of tall claims about evolution: that it can become uncontroversial; that the basic principles are easy to learn; that everyone should *want* to learn them, once their implications are understood; that evolution and religion, those old enemies who currently occupy opposite corners of human thought, can be brought harmoniously together.

Can these claims possibly be true? Isn't evolution the most controversial theory the world has ever seen? Since it's a scientific subject, isn't it hard to learn? If the implications are benign, then why all the fear and trembling? And how on earth can the old enemies of evolution and religion do anything other than come out of their opposite corners fighting?

I might be an optimist, but I am not naive. Allow me to introduce myself: I am an evolutionist, which means that I use the principles of evolution to understand the world around me. I would be an evolutionary biologist if I restricted myself to the topics typically associated with biology, but I include all things human along with the rest of life. That makes me an evolutionist without any qualifiers. I and my fellow evolutionists study the length and breadth of creation, from the origin of life to religion. I therefore have a pretty good idea of what people think about evolution, and I can report that the situation is much *worse* than you probably think. Let me show you how bad it is before explaining why I remain confident about accomplishing the objectives of this book.

Most people are familiar with the reluctance of the general public to accept the theory of evolution, especially in the United States of America. According to the most recent Harris Poll, 54 percent of U.S. adults believe that humans did not develop from earlier species. That is *up* from 46 percent in 1994. Rejection of evolution extends to beliefs about the origin of other species, the fossil record as evidence for evolution, and the constant refrain that evolution is "just a theory."

To make matters worse, most people who do accept evolutionary theory don't use it to understand the world around them. For them it's about dinosaurs, fossils, and humans evolving from apes, not the current environment or human condition. The polls don't measure the fraction of people who relate evolution to their daily lives, but it would be minuscule.

It's easy for scientists and intellectuals to smile at the ignorance of religious believers and the general public, but the fact is that they're not much better. The Ivory Tower would be more aptly named the Ivory Archipelago. It consists of hundreds of isolated subjects, each divided into smaller subjects in an almost infinite progression. People are examined less with a microscope than with a kaleidoscope—psychology, anthropology, economics, political science, sociology, history, art, literature, philosophy, gender studies, ethnic studies. Each perspective has its own history and special assumptions. One person's heresy is another's commonplace. With respect to evolution, most scientists and intellectuals would say that they accept Darwin's theory, but many would deny its relevance to human affairs or would blandly acknowledge its relevance without using it themselves in their professional or daily lives. In effect, there is a wall within academia that restricts the study of evolution to biology and a few human-related subjects such as human genetics, physical anthropology, and specialized branches of psychology. Outside this wall, it is possible for a person to get a Ph.D. without a single course in evolution or more than a casual reference to evolution in other courses. That is why the term "evolutionary biologist" sounds familiar while the more general term "evolutionist" has a strange ring.

Some intellectuals rival young-earth creationists in their rejection of evolution when it comes to human affairs. A 1997 article in *The Nation* titled "The New Creationism: Biology Under Attack" put it this way:

> The result is an ideological outlook eerily similar to that of religious creationism. Like their fundamentalist Christian counterparts, the most extreme anti-biologists suggest that humans occupy a status utterly different from and clearly "above" that of all other living beings. And, like the religious fundamentalists, the new academic creationists defend their stance as if all of human dignity—and all hope for the future—were at stake.

The famous metaphor of the mind as a blank slate captures the idea that we can understand the human condition without any reference to basic evolutionary principles or our own evolutionary past. The most extreme academic creationists reject not just evolution but science in general as just another social construction, but they are only one particularly fierce tribe that inhabits the Ivory Archipelago. Other tribes are fully scientific but still manage to exclude evolutionary theory. In a 1979 survey of twenty-four introductory sociology textbooks, *every one* assumed that biological factors were irrelevant to the study of human behavior and society. Fast-forwarding to the present, political scientist Ian Lustick could say this about the human social sciences in a 2005 article:

> Of course social scientists have no objection to applying evolutionary theory in the life sciences—biology, zoology, botany, etc. Nevertheless, the idea of applying evolutionary thinking to social science problems commonly evokes strong negative reactions. In effect, social scientists treat the life sciences as enclosed within a kind of impermeable wall. Inside the wall, evolutionary thinking is deemed capable of producing powerful and astonishing truths. Outside the wall, in the realm of human

behavior, applications of evolutionary thinking are typically treated as irrelevant at best; usually as pernicious, wrong, and downright dangerous.

It might seem that the situation can't get more bleak, but it does. Evolutionary biologists are themselves conflicted about the study of our own species. When Harvard evolutionary biologist Edward O. Wilson published his encyclopedic book *Sociobiology* in 1975, his fiercest critics were fellow Harvard evolutionary biologists Stephen Jay Gould and Richard Lewontin. Fast-forwarding to the present, the National Science Foundation's most recent and ambitious effort to fund evolutionary research is called the National Evolutionary Synthesis Center (NESCent), whose basic mission is to "help foster a grand synthesis of the biological disciplines through the unifying principle of descent with modification." This language is not as grandiose as it might seem. Biologists expect evolution to serve as a unifying theory, delivering "powerful and astonishing truths," as Ian Lustick put it. Yet, as a curious complement to his diagnosis of the social sciences, not a single member of NESCent's scientific advisory board represents a human-related subject apart from human genetics. It seems that the barrier separating the study of humans from the study of the rest of life is largely respected on both sides, even by evolutionary biologists who are trying to foster a grand synthesis.

Knowing all of this, I remain confident that there is a path around both walls of resistance, the first denying evolution altogether and the second denying its relevance to human affairs. Darwin provides an example for us to emulate: on any given day of his life he might have been found dissecting barnacles, minutely observing the behavior of his children, or germinating seeds that had first been fed to mice, which in turn had been fed to hawks at the London Zoo. The same person who studied earthworms and orchids also studied human morality. Darwin's interests were so far-flung that his mail came by the wagonload from all corners of the globe. One letter about plant distributions in India might be followed by another on the emotional expressions of African natives. Darwin's empire of thought was larger than the British Empire.

How was Darwin able to unite so many subjects and blend humans seamlessly with the rest of life? Perhaps he was a genius. Perhaps there was less to know back then. Perhaps, but the main reason is more interesting and relevant to our own situation. It was primarily Darwin's *theory*, not his personal attributes or time and place, that enabled him to build his empire of thought. Moreover, his theory was powerful even in a rudimentary form, because Darwin knew so much less about the details of evolution than we do now.

The same theory enables modern evolutionists to build empires of their own. I'm no Darwin, but my own career shows what a good theory can do. I have studied creatures as diverse as bacteria, beetles, and birds. I have studied topics as diverse as altruism, mating, and the origin of species. I can understand and enjoy the work of my colleagues who study an even greater range of creatures and topics. Please don't think that I am boasting about myself—that would be boring. I am boasting about the *theory*, and the whole point of this book is to show how *anyone* can profit from it. It takes a great theory, not great intelligence, to acquire this kind of synthetic knowledge.

If our own species can be included in this grand synthesis, there is every reason to do so. It would be like a strange figure emerging from the shadows to enjoy the warmth of a campfire with good company. My own career shows that this is possible. Just like Darwin—not because I share his personal attributes but because I share his theory—I have seamlessly added humans to the bestiary of animals that I study, on topics as diverse as altruism, beauty, decision making, gossip, personality, and religion. I publish in anthropology, economic, philosophy, and psychology journals in addition to my biological research. My books are on subjects that most people don't associate with evolution: *Unto Others: The Evolution and Psychology of Unselfish Behavior* (co-written with a world-class philosopher named Elliott Sober), *Darwin's Cathedral: Evolution, Religion, and the Nature of Society,* and *The Literary*

Animal: Evolution and the Nature of Narrative (co-edited with a bold young literary scholar named Jonathan Gottschall). These are not popular accounts watered down for a general audience. They are written for the experts, most of whom spend their lives studying a much smaller range of subjects. Evolutionists can stride across human-related subjects at the highest level of intellectual discourse, in the same way that evolutionary biologists are already accustomed to striding across biological subjects.

Darwin should be emulated in another respect. His interactions with people from all walks of life were primarily respectful and cordial. We can learn from his humility and good humor in presenting his theory to others, in addition to the theory itself. Since writing *Unto Others* and *Darwin's Cathedral*, I have spoken about evolution, morality, and religion to diverse audiences around the world. Perhaps my most memorable experience was a televised conversation with a group of faculty and monks from St. John's University in Minnesota, a Catholic university and the oldest Benedictine monastery in North America. My co-author Elliott Sober was invited to converse with His Holiness the Dalai Lama, making me unspeakably jealous. These encounters are the very opposite of the sterile "debates" that are staged between creationists and evolutionists. If this kind of cordial dialogue can take place for evolution and religion, then surely it can take place for evolution and any other human-related topic.

Indeed, evolution is increasingly being used to study all things human in addition to the rest of life. I recently conducted an analysis of the highly respected scientific journal *Behavioral and Brain Sciences* (*BBS*). The format of *BBS* is for a lengthy target article to be followed by commentaries by other authors, providing a comprehensive exploration of a particular topic. The topics covered by *BBS* are highly diverse, from neuroscience to cultural anthropology. *BBS* articles are subject to a bruising review process before they are accepted, not to speak of the scrutiny they receive in the commentaries. They often have a large impact on subsequent research. According to a formula that the scientific publishing industry uses to calculate the impact of its journals, *BBS* is ranked first among forty behavioral science journals and seventh among 198 neuroscience journals. If anything qualifies as solid and trendsetting science, it is a *BBS* target article.

My analysis shows that during the period 2000–2004, 31.5 percent of the *BBS* target articles used the word "evolution" in the title or as a keyword, for topics as diverse as religion, schizophrenia, infant crying, language, food transfer in hunter-gatherer societies, facial expression, empathy, vision, brain evolution, decision-making, phobias, mating, cultural evolution, and dreams. In other words, *using evolution to study our own species is not a future event or fringe science. It has already arrived.*

Curious, I e-mailed a survey to the authors of these target articles to learn how they acquired their evolutionary expertise. I discovered that most of them received their formal training in other areas (such as psychology, anthropology, or linguistics) with little or no exposure to evolution in college or graduate school, as might be expected from my bleak portrayal of the Ivory Archipelago. Instead, they encountered evolutionary theory on their own, often by happenstance, and gradually built up their expertise until it became a guiding force in their research. The fact that they could train themselves so easily indicates that the power of evolutionary thinking resides not in a mass of technical detail but in something much simpler that perhaps anyone can learn. In many ways they recapitulated the experience of Darwin, who built his empire of thought without the benefit of the masses of technical detail available to us today.

Perhaps you can begin to see why I am confident about achieving the objectives of this book, but it gets better. I am a teacher in addition to a scientist. Every year I teach a course called "Evolution for Everyone" that is open to all students. Last year they came from the following departments: anthropology, art, biology, business, chemistry, cinema, computer science, creative writing, economics, education, engineering, English, history, human development, linguistics, management, mathematics, nursing, philosophy, physics, political science, and psychology. They included new students fresh from high school, seasoned upperclassmen, and even more seasoned adults from the community continuing their education. At the end of the semester the students are asked to evaluate the class anonymously, so they have nothing to gain or lose by their comments. Here is a sample of what they have to say:

> *"This course provides evidence that evolution is evident in everything. It revolutionized my way of viewing problems."*
>
> *"This course changed the way I look at things in general. I try to see and understand them from an evolutionary perspective now."*
>
> *"This course shows how positive evolution can be. It teaches you a great deal and increases one's interest."*
>
> *"I came into the class not knowing a lot about evolution. I now have an entirely new outlook on how evolution can be applied to many aspects of life."*
>
> *"I had taken evolution classes in high school and never felt it was interesting. But this class changed all my views upside down. The most boring theories became interesting thoughts that I could relate to in my own day to day life."*

I wish that I could attribute these glowing comments to my prowess as a teacher, but once again the credit goes to the theory. My contribution was to help the students recapitulate the experience of Darwin, much as the *BBS* authors did on their own initiative. To learn more about how my students responded to the course, I gave them a survey that measured their political and religious values, science background, prior knowledge of evolution, and general thinking skills. The details will be published in a technical journal for anyone to scrutinize my methods, but here are the results:

- The *majority* of students, not just a select few, learned to think about evolution as a powerful way to understand the world in general and especially their own interests and concerns.

- A background in science or prior knowledge of evolution was *not required*. Freshman English majors got
the message just as strongly as senior biology majors.

- The course succeeded across the *entire range* of political and religious beliefs, from feminists to young
Republicans and from atheists to believers. It might
seem incredible that a religious believer can think
positively about evolution, not with difficulty but just
as easily as anyone else, but this fact will make more
sense as we proceed.

- The students increased their *general thinking skills*, not
just their knowledge about evolution. In plain language, they got smarter. It might seem incredible that
learning about evolution can make you more intelligent, but think about it: Darwin built his empire of
thought by applying a single set of principles to a vast
diversity of subjects, which comes close to a definition
of general intelligence.

For many students who take a course such as the one that I
offer, learning about evolution is like walking through a door
and not wanting to return. Using it to think about their interests and concerns becomes second nature, like riding a bicycle. They are eager to develop their expertise in subsequent
courses and disappointed by professors who do not share
their newfound perspective. In response to this demand, I
and my colleagues at Binghamton University created a program called EvoS that enables anyone to use evolutionary
theory to explore the pageant of life on earth, including the
pageant of human life (http://bingweb.binghamton.edu/
~evos; all major Web sites described in this book are also
listed on p. 381–2). I like to think of EvoS as a new island in
the Ivory Archipelago, a tropical paradise, so to speak. A student describes it as well as I can: "EvoS provides a stimulating
atmosphere within which biologists, psychologists, anthropologists, philosophers, social scientists, and even those in
the arts can transcend traditional academic boundaries and

collaborate in addressing mutually interesting questions. It creates a think-tank atmosphere of sorts, and it's a beautiful thing!"

Now I hope you can see why I am confident about accomplishing the objectives of this book. In a sense they have already been accomplished. There is already a path around the two walls of resistance, the first denying evolution altogether and the second denying its relevance to human affairs. It is not a narrow path accessible to only a courageous few but a thoroughfare regularly traveled by many. In this book I have tried to provide the barest of essentials so that you, the reader, can begin to travel the same thoroughfare. Do not be misled by the simplicity of the chapters. Simplicity is a virtue, and the chapters cover the same issues discussed by evolutionists at the most advanced level. I hope you will agree with me by the end that when it comes to evolution and its widespread acceptance, the future can differ from the past.

Clearing the Deck

EVOLUTIONARY THEORY IS A ship that has weathered many storms. Lately it has been buffeted by the winds of creationism and its born-again cousin, intelligent design (ID), but the enemies of evolution are tame compared to its friends. Herbert Spencer, an intellectual giant of Darwin's day, liked evolution because he thought it justified the inequalities of British class society. The first half of Ronald Fisher's classic *The Genetical Theory of Natural Selection* is still read today, but the second half, on eugenics (the improvement of society by selective breeding), is politely ignored. Hitler liked evolution because he thought it justified the ultimate social inequality of genocide. Using evolution to justify social inequality has become known as "social Darwinism." Darwin himself was passionately against slavery and thought that social policy should be based on compassion, which he regarded as "the noblest part of our nature," but even he was not a New Age guy by modern standards. If we judged Darwin's theory by some of the company it has kept, especially during its early days, we would give it a wide berth.

I have an unromantic view of science and scientists. I regard science as a roll-up-your-sleeves activity, like gardening or construction work. I regard scientists as just like other

folks. No one can be trusted on the basis of their job title—
not scientists, politicians, priests, or self-righteous intellectuals. Trust requires accountability. Some individuals are
accountable on their own, but that's not good enough at the
institutional level. An effective government, religion, or scientific culture must include mechanisms that make everyone
accountable. When I was immersing myself in the study of religion to write *Darwin's Cathedral*, I discovered that the
strongest and most effective religions owe their success in
part to their accountability. As one example, the church
started by John Calvin in sixteenth-century Geneva was run
not by one person but by a group of pastors. When they
couldn't agree, the decision-making circle was widened, not
narrowed. Elders assigned by the church to oversee different
sections of the city had to be approved by those who were
being overseen. Life-threatening jobs such as ministering to
the plague hospital were assigned by lottery. Double book-
keeping methods were used to keep track of charitable funds.
Calvin's church was trustworthy as an institution because it
included mechanisms that made it difficult for even the leaders to cheat—just what you'd expect from a religion founded
on the doctrine of original sin!

Science is largely a way to ensure accountability for factual
claims. In the last chapter I made some factual claims about
how students responded to my "Evolution for Everyone"
course. I am expected to provide the details of my survey to
anyone who wishes to scrutinize them. I abide by an honor
code (never, ever, deliberately falsify your data) that is largely
maintained and when broken results in punishment and exclusion. People can freely disagree about the veracity or interpretation of my results. They can reserve judgment until the
study is repeated or the same claim is demonstrated using different methods. When the process works well, it results in
the accumulation of factual claims that are *supported*, like a
sturdy scaffold that we can stand upon with confidence to
build higher. But vigilance is required at all times, just as with
priests and politicians.

This practical view of science can help us clear the deck of
past controversies about evolution to prepare for our own

voyage. With respect to Hitler and others who used evolution to advance their nefarious causes, it's not as if the world was a nice place before Darwin and became nasty on the basis of his theory. America was colonized before Darwin and the pioneers used the principle of divine right to dispossess the natives. While we're at it, let's not romanticize the natives. In many indigenous cultures, the word for "our people" is "human" and outsiders are classified as a type of animal. Before Darwin, religious tracts claimed that Negroes didn't have souls and women were designed by "God and Nature" for a life of domestic servitude. Is it a surprise that the theory of evolution would be pressed into the same kind of service? Is that a reason to reject the theory?

Science, religion, and politics all face the same problem. Some individuals are driven to benefit themselves at the expense of others or their society as a whole. At a larger scale, some groups are driven to collectively benefit themselves at the expense of other groups. In science this problem takes the form of self-serving factual claims. I say that something is true, not because it really is true but because it serves my interests. The most important wrong ideas in science are not just bad guesses; they systematically serve the interests of their proponents. One hundred years ago, respectable doctors and scientists thought that women shouldn't go to college because brain development interferes with ovarian development. This wasn't just a bad guess; it was a bad guess that supported deeply held notions about the role of women in society. An anthropologist friend who studies ethnic conflict tells me that some Catholics in Northern Ireland think they can identify Protestants by their closely spaced eyes. Some Protestants have exactly the same theory about Catholics. These aren't just bad guesses about facial morphology; they're bad guesses that dehumanize the enemy.

To make matters worse, our hidden agendas need not be conscious. It's not as if we see the world clearly and then willfully distort it to serve our purposes. The world we see clearly has already been distorted by unconscious mental processes. Believers in the eye theory didn't make it up, knowing it to be false. They really believe it to be true. Fortunately, even if we cannot always detect our own biases, we can hold each other accountable for our factual claims by scientific methods. It turns out to be just plain wrong that college women have shriveled ovaries. The eye theory can be disproven in a single afternoon with a pair of calipers. If people can't be held accountable for their factual claims at such a basic level, there is little hope of agreeing on larger issues.

How should we respond to someone who uses a scientific theory to justify acts that we regard as reprehensible? There are two lines of defense. One is to question the alleged facts. The other is to question whether the facts justify the acts. Even if going to college did cause ovaries to shrivel, we might decide that women have the right to choose for themselves which organ to develop.

A third line of defense is to declare the theory off-limits. This is tempting in an impulsive sort of way but makes no sense upon reflection. Imagine that someone runs into your house shouting, "There's a horrible monster coming our way! Let's pretend that it has nothing to do with us!" That's what someone is doing when they declare a theory off-limits because they think it might lead to bad consequences. We need to pluck up the courage to determine if the factual claims are wrong (usually because they serve someone's interests) or if they are right, in which case we must decide what to do about them in a morally responsible fashion.

Not all factual claims are contentious. Sometimes there is everything to gain and nothing to lose by knowing the truth of the matter, such as what it takes to keep an airplane in the air. Many facts revealed by evolutionary theory are of this type, and anyone would be a fool to ignore them. However, even the most contentious claims associated with evolution are less threatening than you might imagine when you roll up your sleeves and get to work on them. Self-serving claims usually make bad scientific theories, which can be rejected easily enough upon close inspection. In my class and this book there are no topics that are off-limits. My students are not threatened or offended, and I'll bet that you won't be either.

Darwin's theory is nearing its 150th birthday, and it might

seem that the most important discoveries have already been made. Actually, this is a *short* time and so much remains to be discovered in the future. The science of genetics didn't get started until the rediscovery of Mendel and his peas in 1904. At first "Mendelism" was regarded as an alternative to "Darwinism" and they didn't come together until 1918, thanks in part to the aforementioned Ronald Fisher. DNA wasn't discovered to be the code of life until 1953. Using evolution to think about organisms in their natural environments didn't really gather steam until the 1960s.

Even more remains to be discovered about our own species. For millennia we have imagined ourselves as set apart from the rest of nature. Our thoughts about ourselves are especially prone to self-serving biases. Our academic traditions have developed without reference to evolution for over a century. In anthropology, one of the great questions of Darwin's day was why some people had become "civilized" while others remained in the state of "savagery." The modern approach to this question is illustrated by Jared Diamond's magnificent *Guns, Germs, and Steel*, but back then it seemed reasonable to imagine a linear evolutionary progression that all cultures travel along, with some more advanced than others. In reaction, other anthropologists declared themselves "anti-evolutionists" and insisted that every culture should be studied and appreciated on its own terms, a tradition that persists to this day. In psychology, evolutionary theory was eclipsed by a tradition known as behaviorism, which claimed that people could be shaped in almost any way by learning. If the mind is a blank slate, then who cares about what happened during the Stone Age? The field of economics has been dominated by something called rational choice theory, which claims that people attempt to maximize self-interested values. The values are flexible and economists don't care much about what they might be, as long as they can employ their optimization models. Evolutionary theory was forced into the shadows while these other theories occupied center stage. Even the most talented and open-minded scientists in these fields are handicapped by events that took place before they were born and became the basis of their disciplinary training, as we saw in the case of the *BBS* authors who had to discover evolution on their own.

A theory is merely a way of organizing ideas that seems to make sense of the world. Scientific methods are merely ways of rejecting or supporting factual claims that emerge from theories. We do not automatically perceive the world as it really is, and we are especially prone to self-serving biases. These biases are advantageous for some people in the short run by definition, but they are often harmful to other people and even to everyone in the long run. Agreeing to establish, protect, and abide by a body of factual information therefore has a moral quality similar to the norms of a religion or a democratic government. It's not always easy, especially when one's own interests appear to be at stake, but facts never lead to actions all by themselves. They can only inform a system of values. I would rather live in a society based on good facts interpreted by a good value system than in any other kind of society.

So, what is this theory that seems to make such sense of the world and might even enhance our understanding of ourselves? Now that we have cleared the deck of past controversies, let's raise the sails and begin our own voyage.

3

A Third Way of Thinking

DARWIN'S THEORY OF NATURAL selection is like a recipe with three ingredients. We start with *variation*. Individuals such as you and I differ in just about anything that can be measured, such as height, eye color, or quickness to anger. Then we add *consequences*. The differences between you and me sometimes *make* a difference in our ability to survive and reproduce. Perhaps your superior size enables you to take my stuff or even kill me directly. Perhaps my inferior size enables me to survive the winter on less food. The details depend upon our particular traits and the environments we inhabit. The final ingredient, a sort of yeast that makes the recipe come to life, is *heredity*. For many traits, offspring tend to resemble their parents. Darwin didn't know how heredity worked in a mechanistic sense, but it was a very well-established fact.

When these ingredients are combined, they lead to a seemingly inevitable outcome. As an example, imagine a population of moths that vary in their coloration (first ingredient). Some are more easily spotted and eaten by predators, removing them from the population (second ingredient). The offspring of the survivors resemble their parents (third ingredient), so the entire offspring generation is more cryptic, on average, than the parental generation. If we repeat the process over many generations, and if nothing else happens to complicate our story, the moths will become very hard for predators to spot. If we could print a picture of each generation on a page and use the stack of pages like a flipbook, the moths would seem to melt into their background. They have acquired a trait that helps them to survive in their environment. In evolutionary parlance, they have increased their *fitness* and become *well adapted*.

Is that all there is? Just about. Learning about natural selection is like having a premature orgasm. You think it will take a long time and lead to a tremendous climax, but then it's over almost as soon as it began!

The main question about natural selection is not "What is it?" or "Does it happen?" but "Why is it such a big deal?" To answer this question, imagine that I place an object in your hand, perhaps a glittery rock or a furry mouse, and ask you to explain how it obtained its properties. Before Darwin you would have had only two options. You could say that God designed it according to His intentions. Perhaps the rock is glittery to please the eye and the mouse is a pest to teach us humility. Or you could dismantle it and explain the whole as a product of its parts. Perhaps the rock is glittery because it has quartz crystals and the mouse's fur is made of keratin. The big deal about natural selection is that it provides a third way of explaining the properties of the mouse, although not the rock. You could say that the mouse is endowed by natural selection with the properties that enable it to survive and reproduce in its environment.

It is obvious that the natural selection explanation differs from the theological explanation. The fact that it also differs from the material explanation is a bit more subtle. Suppose that I place a clay sculpture in your hand and asked you how it obtained its properties. You would spend most of your time talking about the shape imposed by the artist, not the properties of clay. The clay *permits but does not cause* the properties of the sculpture. In just the same way, heritable variation turns organisms into a kind of living clay that can be molded by natural selection. Evolutionists such as myself routinely predict the properties of organisms without any reference to what they are made of, not even their genes.

To illustrate the power of this way of thinking, consider the subject of infanticide. Since natural selection is all about having offspring, it might seem pathological to kill them. Nevertheless, with a little effort you might be able to think of some environmental situations that make infanticide adaptive. In my class I have the students turn to their neighbors to discuss possibilities for a few minutes before telling me their suggestions. They reliably identify the following situations that make infanticide adaptive.

The first is *lack of resources*. If a mother can't feed herself and care for her existing offspring, then it will not increase her fitness to have new offspring. The second is *poor offspring quality*. It will not increase the fitness of a parent to raise an offspring that itself is unlikely to survive and reproduce. The third is *uncertain parentage*. With some important exceptions (read on!), it will not increase the fitness of a parent to raise someone else's offspring. This third situation is confronted primarily by males in mammals such as ourselves, but a female bird can also be "duped" by another female who lays eggs in her nest.

My students (and perhaps you) did very well, since these are indeed the "big three" situations that favor the evolution of infanticide, according to scientists who actually study the subject. The message of this exercise is simple but profound: *How did my students become so smart?* They hadn't read anything on infanticide, and I certainly hope that they hadn't experienced it for themselves. Their evolutionary training had only just begun, but even a tiny bit enabled them to become experts, honing in like heat-seeking missiles on the predictions that are made by the experts. That is the power of natural selection thinking that makes it such a big deal.

My students had also become taxonomic experts. They didn't need to be entomologists, ichthyologists, herpetologists, ornithologists, or mammalogists to say something important about insects, fish, reptiles, birds, and mammals. Natural selection thinking is based on the relationship between an organism and its environment, regardless of its taxonomic identity.

Even better, my students became experts *without any*

knowledge of the physical makeup of organisms. In any particular species, infanticidal behavior will be caused by physical mechanisms that work through certain pathways, but we don't need to know this to predict the occurrence of the behavior. As long as we have the living clay of heritable variation, we can make predictions based on the shaping influence of natural selection.

Finally, *we can repeat the exercise to become expert on an infinite number of other subjects.* Why are males larger than females in some species and the reverse in others? Why are there two sexes in the first place? Why are males and females born in equal proportions in some species but not others? Why do some organisms reproduce once and then die, while others reproduce at repeated intervals? Why do some plants live three weeks and others three thousand years? Why are some organisms social and others solitary? Among social organisms, why do some individuals cooperate and others cheat? Questions such as these are truly without end.

Soon I will add some qualifiers to the power of natural selection thinking, but for now we should pause to appreciate its amazing explanatory scope. I began this book by saying that just about anyone can become an evolutionist by learning to think like Darwin. This might have seemed crazy, but now I hope you can see that it is a very sensible and down-to-earth suggestion. It is the core idea of natural selection that we need to learn, and it is powerful even in its most rudimentary form. This was true even for Darwin, who recalled his moment of insight as if struck by a bolt of lightning.

> I can remember the very spot in the road, whilst in my carriage, when to my joy the solution occurred to me. . . . The solution, as I believe, is that the modified offspring of all dominant and increasing forms tend to become adapted to many and highly diversified places in the economy of nature.

Before this insight, Darwin had been struggling with thousands of bits of information that refused to fit together. Then he saw how they could become one big picture like a jigsaw puzzle. No wonder that he used the word "joy" to describe the experience!

I also experience joy as I survey the length and breadth of creation, dropping in on subjects and organisms that strike my fancy: shyness and boldness in fish on Monday, egg laying in birds on Tuesday, human gossip on Wednesday. It's a joy just to think about such different things as part of a single picture, but it's a special delight to work on them at the professional level, discovering new pieces and how they fit together for the first time. How amazing that virtually all of us can begin to expand our horizons and fit the pieces of our world together on the basis of such a simple idea.

4

Prove It!

EVOLUTION IS A GREAT theory for predicting the properties of organisms, but no theory is guaranteed to be correct. Unless the predictions can be tested by scientific methods, progress grinds to a halt. Depending upon who has your ear, you might think that evolutionary theory remains unproven after all these years. Even someone as distinguished as the president of the United States (George W. Bush) has declared that the jury is still out. Let me show you how silly this is, once you take a roll-up-your-sleeves approach to science.

I have studied infanticide in a wonderful insect called the burying beetle (family Silphidae, genus *Nicrophorus*). If you want to see a burying beetle for yourself, take a piece of raw meat about the size of an egg and tie a long piece of dental floss tightly around it. Place it on the ground in a wooded area and come back the next day. There is a good chance that the meat will be gone but the dental floss will still be visible. Follow the dental floss to where it disappears into the ground, remove the soil gently with a spoon, and you will probably find the meat with two large beetles in Halloween colors, orange and black. Look closer and you will see mites on their backs like passengers on an airplane.

Burying beetles are carrion feeders that specialize on small carcasses, such as mice and baby birds that fall out of their nests. They are one of very few insects that practice family values: both the mother and the father stay to raise their offspring. Working together, they drag the carcass to a suitable spot and bury it in an underground chamber for protection. This by itself is a remarkable feat because the carcass can be many times their own weight and the soil contains impediments such as roots and stones. Once the carcass is buried, the female lays her eggs in the surrounding soil and the couple sets about preparing the carcass for their brood. This involves removing the fur or feathers, rolling the carcass around and around in the crypt until it assumes a spherical shape, and covering it with a secretion that inhibits microbial growth. The parents must tend the carcass continuously because if you remove them, a beard of fungus and bacteria grows almost at once.

When the eggs hatch, the young larvae are actually called to the carcass by the parents, who make a sound by rubbing their hard wing covers against their abdomens. Rather than feeding upon the carcass directly, the larvae cluster around their parents, who regurgitate predigested food. When they become larger the larvae feed on the carcass itself, reducing it to bones while they become fat white grubs. As soon as the larvae tunnel into the soil to transform into adults, the parents fly off separately in search of other carcasses.

This might seem like a heartwarming tale, but there is a dark side. The carcass can vary in size from a baby mouse to an adult squirrel. Do the burying beetles adjust the size of their brood to the size of the carcass, and if so, how is it accomplished? It turns out that they do make the adjustment, and it is accomplished by infanticide. At the same time that they are nurturing some of their offspring, they are munching on others until they have reduced their brood to a size appropriate for the carcass. This illustrates the first environmental situation that favors the evolution of infanticide: lack of resources.

How can I state these facts with such confidence? I went to Kmart and purchased a lot of plastic shoe boxes and potting soil. I filled the shoe boxes with soil and placed a pair of beetles and a dead mouse in each one. Half of the mice were small and the other half were large. I dismantled a third of the boxes after the eggs had been laid and before they had hatched. I counted the eggs and discovered that females who bury small mice lay just as many eggs as females who bury large mice. I dismantled another third during the larval stage and discovered that brood size had become proportionate to carcass size. I allowed the offspring to complete their development in the last third and discovered that there was no difference in the size of beetles grown on small and large carcasses. In other words, offspring *number* had been regulated with remarkable precision such that average offspring *size* remained constant.

The regulation takes place after the egg stage. How is it accomplished? It might be infanticide, offspring competition, or conceivably even suicide. Direct observation settled the issue. Adults can be easily seen nurturing some offspring and munching on others until brood size is proportionate to carcass size. Their adaptive strategy is to overproduce at the egg stage and cut back at the larval stage. Infanticide is as much a part of this strategy as parental care.

Another question that interested me was why both parents remain to take care of their young. There are many possibilities. After all, being a burying beetle is hard work. You have to drag something that can be several hundred times larger than yourself, bury it, work tirelessly to prevent microbial growth, and so on. I and my colleagues who study this marvelous creature did numerous experiments of the sort that I have described for brood regulation, in both the laboratory and their natural habitat. Remarkably, we discovered that "single moms" did just as well as "intact families" in most respects. We didn't expect this and I still don't entirely believe it, but that is what the results indicate. The main situation that requires Dad in addition to Mom involves repelling intruders.

It turns out that of all the hardships faced by burying beetles, the most dangerous is other burying beetles. With their exquisite sense of smell, they can detect a buried carcass, and when they find one a battle ensues. Males and females are equally fierce, and there is no average difference in size. If the intruder wins, he or she kills the offspring to make room for his or her own. The resident member of the same sex is expelled and the resident member of the opposite sex dutifully mates with the killer of his or her children. This is the third environmental situation that favors the evolution of infanticide: when the infants are not your own.

How about the second environmental situation that favors the evolution of infanticide—poor offspring quality? If the parents are reducing the size of their brood, wouldn't it make sense for them to assess their quality to retain the best? I can't answer this question because the appropriate experiment has not yet been done. However, I'll bet you can imagine how it might be designed. Go to Kmart and get more shoe boxes and potting soil. In half of the boxes, intercept the young as they are arriving at the carcass and reduce the brood size yourself by removing some at random. Let the parents do it their way in the other half of the boxes. Compare the quality of the offspring that emerge (their ability to survive and reproduce) and celebrate your newly documented fact with a beer.

This roll-up-your-sleeves kind of science documents facts the way a brick factory produces bricks. A brick is a humble object, of little use by itself, but it is durable and becomes great in combination with other bricks. So it is with the facts documented by scientific methods. A fact such as "burying beetles regulate brood size but not at the egg stage" is humble and amounts to little by itself. However, assuming that it survives the scrutiny of other scientists, it is durable and will be as true five hundred years from now as it is today. The idea that science is hard to do and facts can never become reliable is incorrect at the brick level.

Big theories such as evolution and creationism are evaluated not by a single decisive experiment but by how well they interpret the countless facts that lie around us like piles of bricks. Creationism fails as a theory in part because it is so unhelpful. Does anyone know God's will for burying beetles? Natural selection thinking, by contrast, delivers very specific predictions, such as "infanticide should be expected primarily in three situations." The predictions might be right or wrong, but at least they tell us what to look for. When a detailed prediction is confirmed, it provides support for the theory that issued the prediction—never decisive, only cumulative. You might not be convinced about natural selection writ large, but you must admit that it did a very good job telling us where to find infanticide in burying beetles.

And not just burying beetles. Scientists are busy people and have been churning out durable facts about infanticide in all sorts of species. If you have access to the Internet, go to Google Scholar (http://www.scholar.google.com), a specialized branch of Google that searches the scientific literature. Type "infanticide" and you will get more than 25,000 hits providing information on dozens and dozens of species, from other insects to the mountain gorilla. Overwhelmingly, infanticide occurs in the "big three" situations that my students were able to identify with just an ounce of knowledge about natural selection.

You will also find plenty of information on infanticide in *our* species. Are you curious to know if we stand apart from the rest of nature or if we fit the same pattern? My students are, but I must keep you in suspense until a future chapter.

You might be surprised that a burying beetle with a brain the size of a mustard seed is capable of such sophisticated behavior. Dragging, burying, modifying, assessing—doesn't all of that require *intelligence*? Isn't intelligence something that we have and they don't? Evolutionary theory does not make a strong prediction on this question. At first glance, it is plausible that sophisticated behaviors require big brains. If so, then creatures with small brains would behave in unsophisticated ways (such as laying eggs but not regulating brood size), creatures with big brains would behave in sophisticated ways (such as regulating brood size), and we would be most sophisticated of all. Most evolutionists fifty years ago probably expected this to be the case, but the facts told a different story. It turns out that most species are very, very good at solving the problems relevant to their survival and reproduction, as we have seen in the case of burying beetles. Animal mental abilities can vastly exceed our own for specific tasks. Some ants can find their way home across a featureless landscape by using the sun as a compass. Can you do that? A chickadee can remember the location of thousands of food items that it stored during the fall and recovers during the winter, whereas we have trouble finding our keys. However, we are indeed special in the *flexibility* of our intelligence. Our species can solve brand-new problems in a way that other species cannot.

This fact was elegantly established for burying beetles over a century ago by the great French naturalist Jean-Henri Fabre. To show that burying beetles do not have humanlike intelligence, he tied one end of a wire to a dead mouse and the other end to a stick stuck in the ground. The beetles couldn't chew through the wire, so to bury the mouse they needed to topple the stick by excavating under it. This was not a *difficult* task because the beetles are masters at soil excavation. However, it was a *new* task because the beetles are accustomed to excavating under the mouse. A person would quickly solve the new problem, but the beetles were dumbfounded. They just mechanically kept doing what they were programmed to do, without a glimmer of awareness or creativity.

Fabre was a master experimentalist who established facts such as these by the dozen, as solid today as when they were first documented. He was also a creationist, as most naturalists were at the time, although creationism back then bore little resemblance to its current form. Darwin admired him and called him "that inimitable observer." Darwin knew that facts are common property that can be used by any theory. His theory was so superior at explaining the facts, even way back then, that support for creationism rapidly waned. This was not a conspiracy against religion, since the evolutionists were vastly outnumbered. It was the way the facts fell. Fabre's facts will last forever, but the way he interpreted them evaporated like the morning mist.

What I have shown for infanticide can be repeated for innumerable other subjects. In fact, my primary interest in studying burying beetles was not infanticide or even the beetles. It was the *mites* that you might notice if you look closely at the beetles. They too are specialized carrion feeders but, lacking wings, they require the beetles to travel from carcass to carcass. There can be more mites on a single beetle than passengers on a jumbo jet, but that's another story.

5

Be Careful What You Wish For

NUMBERLESS FOLKTALES TELL the story of someone who is granted a wish, only to have it go horribly, horribly wrong. So it is with natural selection. Nature can be so inspiring: the butterfly's wing, the towering redwood, the incredible machinery of the body and mind. It is easy to imagine that adaptations are always benign, just as the characters of the folktales believe that their wishes will make them live happily ever after. But adaptations, like wishes, are not quite what they seem. Infanticide has already given us a taste of what can go wrong. Picture me sitting down with my family to dinner, only it's not a turkey we are carving but little Bobby, our last-born. There just wasn't enough food to go around this year. Then an enormous woman bursts through the door, chases my wife out of the house, slays my other children in front of my eyes, and expects me to mate with her right away. These are adaptations that evolved by natural selection in burying beetles, but we do not find them inspiring or wish them for ourselves!

The visceral disgust that this tale provokes is not confined to tenderhearted religious believers. One of the first scientists to think seriously about infanticide as an adaptation was the primatologist Sarah Blaffer Hrdy, who observed male langurs (an Asian primate) barging into a troop, chasing away the resident male, and attempting to kill the offspring, just as I described for burying beetles and in my ghoulish story. Her suggestion that infanticide was adaptive for the males was greeted with outrage and disgust by many members of the scientific community. How could such a despicable behavior be *adaptive*? Surely it must be pathological, perhaps caused by something *unnatural*, such as overcrowding. Tempers ran high, but progress was still made thanks to scientific methods. Proponents of the overcrowding theory were forced to abide by predictions about when infanticide might occur. The over-crowding hypothesis wasn't supported by the facts and slowly faded away.

The heart of the matter is that *the evolutionary concept of adaptation frequently departs from what we regard as benign in the everyday sense of the word.* Adaptations can be the epitome of shortsighted selfishness, even harming everyone over the long run. This is deeply threatening because it seems to imply that the world is not and can never be as we wish it, that our utopias are mere comforting illusions. It is common to respond in a number of ways, all of them unsatisfactory.

If you're a cynic, you just smile and say "I told you so." Cynics already believe that our utopias are comforting illusions, so for them the dark side of evolutionary theory does not pose a problem. I am not a cynic, as you have probably guessed by now. I like to accept goodness at face value, and I think that the world can be much better in the future than in the past or present. Moreover, I think I can *prove* that evolutionary theory allows room for my kind of optimism, as strange as that might seem.

Another response is to declare evolution off-limits. I have already described this as like a person shouting, "There's a monster coming! Let's pretend that it has nothing to do with us!" At the very least we should pluck up the courage to face the threat.

A third response is to sanitize evolution by focusing primarily on its unthreatening, awe-inspiring creations. An example is a recent book titled *Inside the Mind of God: Images and Words of Inner Space*, by Sharon Begley and Michael Reagan, which couples stunning images of the microscopic natural world with inspirational passages. To its credit, it in-cludes an informed introduction and stunning images of can-cer cells and the virus that causes AIDS in addition to DNA and nerve synapses, but the goal of the book is clearly to cre-ate the same kind of reverence that one feels when entering a cathedral. There's nothing wrong with that, but it is an enter-prise different from the roll-up-your-sleeves kind of science that I am attempting to convey, which must pay equal atten-tion to the good, the bad, the beautiful, and the ugly.

A fourth response is to face the evolutionary music and de-clare that we can improve life by somehow "rebelling against our genes," as if evolution leaves no room for optimism and some other theory does. My problem with this response is that the other theory is never clearly described. It just arrives on the scene to save the day, like the hero of a bad movie. You might be thinking, "What about learning and culture? Aren't these the heroes that we have always relied upon to save us from our villainous genes?" Right, but later I will show that learning and culture are better understood within the frame-work of evolutionary theory than as alternative theories.

What response *is* acceptable, if these are not? My story has a happy ending, made happier by the fact that it is fully real-istic. Here is how I present it to my class. First I ask them to list the traits that they associate with goodness. How would they describe the morally perfect individual? Then I ask them to list the opposite traits that they associate with pure evil. There is only one catch: just as in a wish-granting folktale, *they are not allowed to change their minds.* My students have fun with this exercise and typically nominate the following traits as good and evil.

Good traits	Evil traits
Altruism	Selfishness
Honesty	Deceit
Love	Hatred
Sacrifice	Avarice
Bravery	Cowardice
Loyalty	Betrayal
Forgiveness	Spite

These lists are unsurprising, the standard portrayal of good and evil. Now I ask them to consider three imaginary experiments:

What will happen if you put a good person and an evil person together on a desert island? My students regard this as a no-brainer. The good person will become shark food within days. As an old Ray Charles song puts it, "She took advantage of my goodness and made a pitiful fool out of me."

What will happen if you put a group of good people on one island and a group of evil people on another island? This is also a no-brainer. The good group will work together to escape the island or turn it into a little paradise, while the evil group will self-destruct.

What will happen if you allow one evil person to paddle over to Virtue Island? The answer to this question is not obvious because it is a messy combination of the easy answers to the first two questions.

The message of this exercise is simple but profound. It shows that *goodness can evolve*, at least when the appropriate conditions are met. Groups of individuals who exhibit good traits are likely to survive and reproduce better than any other kind of group. The problem with goodness is its vulnerability to subversion from within. To the extent that natural selection is based on fitness differences *within* groups, the traits associated with evil are the expected outcome. To the extent that natural selection is based on fitness differences *among* groups, traits associated with goodness are the expected outcome. If you're quick, you might be asking, "Can't behaviors that count as good *within* groups be used for evil purposes *among* groups?" The answer is yes, illustrating another way that adaptations can go bad: groups whose members are as good as gold toward each other, wreaking havoc against other groups. Sound familiar? I will return to this problem (and potential solutions) later, but for now we need to appreciate the progress that we have made: what my students (and perhaps you) regarded as good can evolve as a biological adaptation. The only requirement is to have individuals exhibiting good traits interact with each other and to exclude individuals exhibiting evil traits.

I chose my words carefully in the previous sentence, saying "individuals exhibiting good (or evil) traits" rather than "good (or evil) individuals." This is because we can easily imagine individuals who are flexible in their behavior, capable of exhibiting traits that are good or evil. We can't label flexible individuals, only how they behave, and goodness evolves when the *behaviors* become appropriately clustered.

Some of my students are tempted to change their mind about what counts as good and evil by the end of the example, even though I told them to choose carefully. There is a pervasive tendency, especially in our own culture, to equate "successful" with "selfish." One way of doing this is by imagining what a selfish individual would do, knowing that he will be grouped according to how he behaves. Such an individual would choose to be good in the second experiment, because his only choice is to be a good individual in a good group or an evil individual in an evil group. Doesn't that make goodness selfish, at least whenever it works? The important point—however one wants to think about it—is that there are two very different pathways to evolutionary success. One involves exploiting your neighbor and the other involves working with your neighbor to achieve jointly beneficial outcomes. The second pathway provides room in evolutionary theory for what we call goodness.

Seeing both pathways goes a long way toward reducing the threat that is so often associated with evolution. Earlier I said that evolution is threatening because it seems to imply that the world is not and can never be as we would wish it. Now we can see that this implication is false. If the traits that we associate with goodness can evolve, then we can make them more common by providing the right environmental conditions. Far from denying the potential for change, evolutionary theory can provide a detailed recipe for change.

I also said I could prove that there is room in evolutionary theory for my kind of optimism, which includes accepting goodness at face value and thinking that the world can be better in the future than in the past or present. Although my claim might have seemed extravagant, haven't I done just that? If evolutionary theory leaves room for optimism, is there any need to look elsewhere for something that will "save us from our genes"?

Seeing both pathways also helps us to avoid the errors of sanitizing evolution. The average person knows that immoral behaviors are common and tempting because they benefit the actor, that human groups have a disturbing tendency to restrict goodness to their own members, and so on. In the Old Testament, even God himself commands his chosen people to commit infanticide and smites those who show compassion for the enemy (for example, 1 Samuel, ch. 15). Knowing this about our own species, why should we pretend that evolution or nature as a whole is somehow inherently benign? The towering redwoods are indeed awesome to walk among, but the reason they are so tall is to deprive their shorter neighbors of light. If we must sermonize about nature, let it be a fire-and-brimstone sermon in which goodness can be achieved only by creating the appropriate conditions—by walking the straight and narrow, as a preacher might put it.

Just as we shouldn't portray nature as a cathedral, neither should we portray it as a gladiators' pit. The traits that we associate with goodness and the environmental conditions that enable goodness to evolve are not restricted to our own species. Nature is not good as a whole, but goodness exists in nature in places that you might never expect, as you will shortly discover.

All this talk about good and evil might seem like idle philosophizing and moralizing, far removed from the world of science. On the contrary, if you learn anything from this book, it should be the practical value of evolutionary theory for science as a roll-up-your-sleeves activity. The ideas in this chapter can even be used for something as practical as breeding a better chicken.

It turns out that something very similar to my desert island thought experiment has been performed on chickens by a poultry scientist named William Muir. When I say poultry scientist, don't think of someone in overalls and a straw hat. Poultry science is a sophisticated field, and Bill is very highly trained in genetics and evolution. Chickens have always lived in groups, and in the modern egg production industry they are crammed inhumanely into cages usually containing nine to twelve hens. Bill wanted to increase egg production by selective breeding, and he tried to do it in two ways. The first method involved selecting the most productive hen from each of a number of cages to breed the next generation of hens. The second method involved selecting all the hens from the most productive cages to breed the next generation of hens. You might think that the difference between the two methods is slight and that the first method should work better. After all, it is individuals who lay eggs, so selecting the best individuals directly should be more efficient than selecting the best groups, which might include some individual duds. The results told a completely different story.

When Bill presented his results at a scientific conference that I attended several years ago, he showed a slide of hens selected by the first method after six generations. The audience gasped. Inside the cage were only three hens, not nine, because the other six hens had been murdered. The three survivors had plucked each other during their incessant attacks and were now nearly featherless. Egg production plummeted during the course of the experiment, even though the most productive individuals had been selected each and every generation. What happened? *The most productive individuals had achieved their success by suppressing the productivity of their cagemates.* Bill had selected the meanest hens in each cage and after six generations had produced a nation of psychopaths. The first method corresponded to the first version of my thought experiment, in which a good individual and an evil one are placed together on a single desert island.

Then Bill showed a slide of hens selected by the second method after six generations. The cage contained all nine hens, plump and fully feathered, and judging from their expressions they seemed to be having a good time! Egg production had increased dramatically during the course of this experiment. By selecting whole groups, Bill had selected against aggressive traits and for cooperative traits that enabled the hens to coexist harmoniously. The second method corresponded to the second and third versions of my thought experiment, which includes natural selection at the level of groups. Needless to say, this is the method that the poultry industry has adopted to select for egg productivity in hens.

The slides were so impressive that Bill kindly provided me copies to show in my own talks. After one of my talks a professor ran up to me and exclaimed, "That first slide describes my department! I have *names* for those three chickens!" Evidently her department had adopted a policy of promoting members entirely for their individual accomplishments, with results comparable to the first method of breeding chickens and my first desert island thought experiment. *Genetic* evolution had not taken place, of course, but *something* had taken place that gave comparable results. Soon I will discuss what that "something" might be. For now, whatever we might say about human good and evil, I can tell you with confidence that the eggs in your refrigerator are brought to you by good hens.

6

Monkey Madness

NATURAL SELECTION CAN PRODUCE exquisite adaptations, but this is not invariably so. In some cases they look like Rube Goldberg devices cobbled together out of spare parts. In other cases they fail to evolve at all, even though they would be immensely useful. These facts about evolution are stressed again and again, especially in arguments against creationists and proponents of ID. If you can identify the god or other intelligent agent that created my aching back and wobbly knees, let me know so that I can demand a recall!

The purpose of this chapter is not to hit creationists and ID proponents one more time with the same blunt instrument. In fact, this whole book is on a wavelength that those folks haven't tuned in to yet. To play the roll-up-your-sleeves science game, they would need to use their knowledge of the intelligent designer to predict the properties of organisms in minute detail, such as when burying beetles are likely to commit infanticide. They would need to make their predictions *before* the information is gathered rather than playing Monday-morning quarterback. Only then would they be able to compete with evolutionary theory in the arena of gathering and interpreting facts. They are welcome to try, but I doubt they will succeed. After all, ID proponents can't even tell us if the intelligent designer is a god. Moreover, the very idea of knowing God's will in such minute detail goes against many religious doctrines. According to John Calvin (as I discovered while writing *Darwin's Cathedral*), abandoning the hubris that you can know God's will is the first step toward gaining access to his kingdom.

Critics of evolution aside, imperfection is an important concept for the evolutionist to grasp. My portrayal of natural selection thinking in Chapter 3 was too simple. We can predict the traits that would be adaptive for a given organism in a given environmental situation, but our prediction can fail for a host of reasons. To get a feel for the factors that impede natural selection, I offer the case of the mad monkey.

Our detective is Dr. Stephen Suomi, a senior scientist at the National Institutes of Health who works with primates. As an evolutionist, Steve appreciates the importance of studying animals in relation to their natural environment. Whenever possible he houses his primates in large outdoor enclosures and keeps in close touch with scientists who study the same species in the wild. Every generation, a small fraction of males in his rhesus monkey colony act as if they are out of control. Typical rhesus monkey babies begin life with their mothers and gradually join "play groups" of other monkeys their own age. As they mature, females stay in the same group, while males leave to find their fortunes with other groups. The mad monkeys seemed out of control from the day they were born. Their mothers couldn't handle them. Their friends couldn't handle them. One way that Steve learned to identify them was by the insane risks they were willing to take when leaping from one branch to another. They were literally bouncing off the walls of their enclosure!

(It might seem strange that I am referring to distinguished scientists such as Bill Muir and now Steve Suomi by their first names. For me it is natural, because almost all my interactions with colleagues in far-flung disciplines are on a friendly first-name basis. Besides, when Steve was in college he wore shoulder-length hair and the kind of clothes that comedian Mike Myers parodied in his *Austin Powers* movies. I will be introducing you to many distinguished scientists and scholars in this book and will call them by their first names unless I would feel uncomfortable doing so in real life.)

As a good evolutionist, Steve suspected that this "madness" might actually be biologically adaptive for the individual males, despite the risks they take and the trouble they make for others. His best hypothesis was that the psycho males enter new groups and take over, mate with the females whether they like it or not, and thereby perpetuate their genes. It makes sense but turns out to be false, at least the way the facts have fallen so far. Rejected by family and friends, the psycho males leave their natal groups early and enter new groups before they are ready to compete with the fully adult males. Their reception abroad is the same as at home, forcing them to lead a miserable solitary existence until they die.

Why doesn't natural selection get rid of something so maladaptive? Unable to find a benefit, Steve tried looking for a mechanistic cause. Modern molecular biology enables mechanisms to be studied as never before, and Steve soon had a suspect: a particular version of a gene that influences how the brain responds to the neurochemical serotonin. Just as we tinker with this chemical when we take a drug such as Prozac, the psycho males carried a gene that did the tinkering.

Finding a gene to associate with the behavior was an important step, but in some ways it only deepened the mystery. This gene arose by mutation sometime in the distant past and then *increased* to its current frequency of approximately 10 percent. Individuals carrying the gene survived and reproduced *better* than those without the gene. This can happen by chance for a gene that has no effect on fitness, but Steve had already demonstrated a strong negative effect. Where was the advantage that outweighed the disadvantage, enabling the gene to become reasonably common in the population rather than remaining at a very low frequency?

The solution to the mystery required looking beyond the mad males. The gene is located on one of the autosomes and therefore exists in both sexes. Perhaps its disadvantage in *males* is counterbalanced by an advantage in *females*. Once the gene had been identified, it was easy to check for this possibility. Sure enough, the same gene that produces psycho males also produces confident, capable females who achieve high status within their groups. Moreover, only some of the males who carry the gene become psycho. Other males have the same gene but are fine upstanding monkey citizens. These discoveries would not have been possible without first identifying a candidate gene associated with a particular behavioral syndrome (psycho males) and then looking for the gene in individuals who do not have the syndrome.

Steve then focused on the difference between males who share the same gene but do not share the same syndrome. The main difference turned out to be the parenting style of their mothers. Capable mothers are able to channel the neurochemical systems of their sons to constructive ends, but incapable mothers lose control. This is called an "interaction effect" in evolutionary parlance. A given behavior (such as psycho males) is caused by particular genes interacting with particular elements of the environment in particular ways.

The message of the story is that when we tally the costs and benefits that keep a given gene at a certain level in a population, we need to average across all the different environments experienced by the gene. It will exist within both males and females (unless it is on the y-chromosome in a species such as ourselves). Within each sex it will experience different social and physical environments, which can change over time and space. To make matters more complex, a single gene is not the same thing as a single visible trait. A gene can influence many traits, all contributing to the ledger of costs and benefits. Only if the ledger balances for the *average effect* of a gene will it remain in the population.

We are used to tallying costs and benefits for single organisms. In fact, rational choice theory, which has guided the field of economics for many years, is committed to the notion that individuals are designed to maximize their own personal utilities. This might be true in some cases, and it was reasonable for Steve to look for benefits to balance the costs for his mad male monkeys. But it need not be true in all cases, as Steve learned when he discovered that the costs were concentrated in some individuals and the benefits were concentrated in others.

Evolutionists have known about tallying costs and benefits at the gene level for a long time, which is given the accurate but unexciting term "average effects." In the 1970s, my evolutionist colleague Richard Dawkins wanted to give a more "full-throated" description of evolution for a popular audience, so he coined the new term "selfish gene." That got people's attention but also created a lot of confusion that persists to this day. A gene is called selfish whenever it survives and reproduces better than other genes, all things considered, but that is just newspeak for "anything that evolves." For example, suppose that the traits associated with good and evil in Chapter 5 have a genetic basis. If the good traits evolve because they are appropriately clustered—as in the second chicken experiment and my second and third thought experiments—then Richard would call them selfish because they replaced the evil traits. Never mind that they succeeded by helping each other and despite being exploited by the evil traits that ultimately failed. There is no way for goodness to win without being reclassified as selfish.

With due respect to Richard, we don't need a new term for "anything that evolves," much less a loaded term such as "selfish." If we return to the less shocking but more accurate term "average effect," however, we can see that the gene's-eye view has much to recommend it. It explains how individuals who function poorly can persist indefinitely because they are the unfortunate repository of the costs of genes whose benefits reside in other, more fortunate individuals. On the optimistic side, it shows that the same gene can have very different behavioral manifestations, depending upon the environment with which it is combined. We do not necessarily require "gene therapy" to alleviate monkey madness. We can also employ "environment therapy" by helping mothers to become more capable.

The case of the mad monkey ends with a fascinating twist. Several years ago, Steve started to work with rhesus monkeys that came from China rather than India. His trusted technician, who had been working with the Indian monkeys for many years, soon noticed that the Chinese monkeys were even crazier. Sure enough, the frequency of the gene associated with the psycho males was higher in the Chinese population. Steve has a hunch that the distant ancestors of the Chinese population came from India, which would require crossing the Himalayan Mountains. What kind of monkey would be crazy enough to do something like that?

How the Dog Got Its Curly Tail

DMITRY K. BELYAEV WAS not where he wanted to be. As a Russian geneticist who refused to endorse the ideologically driven theories of his colleague Trofim Lysenko, Belyaev had been removed from a prestigious post in Moscow and sent to Siberia—not to work in the gulag, but to begin a new scientific institute. Even his old job sounds strange to American ears: head of the Department of Fur Animal Breeding at the Central Research Laboratory of Fur Breeding. Nevertheless, an experiment that Belyaev started at his new institute in 1959 revealed something fundamental about evolution in general.

The narrow purpose of the experiment was to create a tamer variety of silver fox. Wild silver foxes had been caged and bred for their fur since the beginning of the twentieth century, but they were by no means fully tame. Belyaev established a rigid protocol for measuring tameness throughout development, from pups a month old to sexual maturity. Individuals were selected for breeding purely on the basis of their tameness index without regard to any other trait. Forty years and forty-five thousand foxes later, Belyaev (who died in 1985) and his successors had produced a new breed that qualified for the exalted title of "domesticated elite." Rather than fleeing from people and biting when handled as their ancestors had, the new breed was eager for human contact, whimpering, sniffing, and licking like dogs, even as little pups less than one month old.

Such is the power of artificial selection, which partially inspired Darwin's theory of natural selection. The most amazing thing about the experiment, however, was that the foxes had become like dogs in *other* respects. Their tails had become curly. Their ears had become floppy. Their coat color had become spotted. Their legs had become shorter and their skulls more broad. These physical traits had not been selected, but they still came along for the ride, as if connected to the behavioral trait of tameness by invisible strings.

These hidden connections force me to amend my description of heritable variation as a kind of living clay that can be shaped by natural selection. That was too simple. Real clay is so malleable that you can shape one part of a sculpture without altering other parts. Living clay is more complex and interconnected, imposing a shape of its own. As a good evolutionist, you might become interested in a conspicuous trait, such as a curly tail or a floppy ear, and ask how it evolved by contributing to survival and reproduction. That's a good question, and I could provide many examples of traits that initially appear as whimsical as a floppy ear but turn out to be important adaptations upon further study. However, the fox experiment reveals a completely different possibility, that the trait has no function whatsoever. It exists because of hidden connections with other traits that can be as obscure as the connection between a curly tail and tameness.

To learn more, we must understand the living clay of heritable variation in more detail. In Chapter 3, I described three ways of thinking based on theology, materialism, and natural selection. Theology is no longer used to explain the material world, not because it is unfairly excluded but because it proved its inadequacy many times over. I mean no disrespect by saying this and will turn my attention to theology in future chapters. For the moment, suffice it to say that the proper business of theology is to establish values, not facts about the material world.

That leaves the other two ways of thinking, based on materialism and natural selection. These are complementary and one can never substitute for the other. Complete knowledge about the physical makeup of organisms will not tell us about the shaping influence of natural selection. Complete knowledge about natural selection will not tell us about the actual mechanisms that evolve or the mysterious hidden connections such as those revealed by the fox experiment. Evolutionists use the terms "ultimate" (for explanations based on natural selection) and "proximate" (for materialist explanations) to make this distinction. Using these terms, we need to learn more about the proximate mechanisms that cause a fox to be wild or tame.

Almost all wild animals are tamer as infants than as adults. This is probably obvious from your own experience and makes good functional sense. A baby fox is incapable of defending itself and lives in a nurturing world provided by its parents. Only later will it have both the means and reasons to run or fight. This adaptive logic (the ultimate explanation) is implemented in real baby foxes by hormonal mechanisms (the proximate explanation). They are born tame, and fear falls like a curtain between two and four months of age, caused by surging levels of corticosteroids, a hormone associated with stress. Their bodies are simply programmed to do this; no experience is necessary. It's easy to imagine a fox baby that must experience a fearful event before becoming fearful, but those babies didn't survive as well as the ones who automatically became fearful at a certain age.

Individual foxes differ in the exact time that the hormonal curtain falls, and this variation is heritable. When Belyaev was selecting foxes purely for their tameness, he was selecting those with the most prolonged infant period. But hormones don't influence single behaviors; they are like conductors that influence many behaviors in a coordinated fashion. The tame foxes remained infantile in many respects, including their floppy puppy ears, short legs, and broad skulls. Another fact of mammalian development is that pigment cells (melanocytes) originate in one part of the embryo (the neural crest) and migrate to their final location on the skin. Prolonging infant development delays their migration and causes some to die, resulting in patches of unpigmented skin—spots.

These are very general rules of mammalian development, which explains a striking pattern. Have you ever noticed that all domestic mammals, including dogs, cats, cattle, horses, pigs, and guinea pigs, tend to share certain traits? They even tend to share the same star-shaped white patch on their foreheads. The ever-observant Darwin noted that "not a single domestic animal can be named which has not in some country drooping ears." Now *there's* a fact that is humble by itself but becomes great in combination with other facts! Among wild mammals, most infants have drooping ears but only the elephant has drooping ears as an adult. *All* of our domesticated animals have become tame by retaining their juvenile traits, just like the foxes in Belyaev's experiment.

Belyaev was a great evolutionist and scientist who struggled against enormous odds. His colleague Trofim Lysenko robbed him of his prestigious post and virtually outlawed the study of genetics and evolution in Russia for decades. His government didn't understand the value of basic knowledge, forcing him to justify his work in terms of the narrow benefits of the fur trade. His nation was so poor that it offered meager support for any kind of science, basic or applied. It's amazing that he could maintain his commitment to basic science under such circumstances.

Compared to Belyaev, a young American scientist named Douglas Emlen is a lucky, lucky man. He is exactly where he wants to be, which is the University of Montana at Missoula. He is doing exactly what he wants to do, which is studying evolution and development, just like Belyaev. Even better, he can concentrate entirely on advancing basic knowledge, which has led him to study a group of beetles without any economic significance but with great scientific significance.

These beetles are a bit like my burying beetles except that they bury dung rather than carrion. Most people who visit Africa marvel at the big game, but another spectacle takes place every time one of those big animals takes a poop. Almost before it hits the ground, hundreds of beetles converge from every direction. I know from experience that you can hear them coming before you can get your pants up. They include dozens of species, from the size of a golf ball to

the size of a lentil, and many of them have horns like miniature flying rhinoceroses.

Doug studies a single genus of dung beetles called
Onthophagus. A genus is the next taxonomic unit above
species, so all its members are derived from a recent common
ancestor. The genus *Onthophagus* includes over two thousand
species, distributed worldwide, and their horn morphology
is amazingly diverse. Some are hornless, while others sport
huge tusks. The horns can exist on their snout, forehead, or
the shield (called a pronotum) behind their head. Molecular
studies have shown that any given horn type within the
genus has evolved not once but numerous times. If we could
turn the evolution of the whole genus into a flipbook, we
would see horns madly sprouting, shrinking, and changing
position. Doug chose the genus because it affords so many
comparisons and his goal is to study the evolution and development of the horns from both an ultimate and proximate
perspective. Of course, the main interest is not in the beetles
per se but what they can tell us about evolution and development in general.

On the ultimate side, we want to know if the horns are
adaptations that evolved by natural selection. As we have
seen, the answer need not be yes. Horns could be like floppy
fox ears that have no function. Once you enter the lives of
these wonderful creatures, however, the function of horns becomes clear enough. Females, who do not have horns in most
species, pack dung into subterranean tunnels to raise their
broods. Males, who do have horns, guard the entrances of the
tunnels against other males. A Homeric battle among males
for females and resources rages under every pile of poop.

Some males are too small to play the warrior game. The
reason that they are small is because they were not provi-
sioned with enough food. When you're a baby dung beetle,
you take what you get and transform into an adult when the
food runs out. It's better to be a wimpy adult than no adult at
all. Wimpy males aren't strong, but they are devious, gaining
access to the females by digging side tunnels of their own.
Horns would be an impediment for this "sneaky" strategy, and
small males don't have horns.

As with the burying beetles that I described earlier, you might be amazed that insects are capable of so many "intelligent" strategies. Not only do males and females engage in very different tasks, but males adopt different strategies as adults depending upon the amount of food they received as larvae. They change not only their behaviors but their very bodies, such as the presence and absence of horns. Dung beetles do not possess humanlike intelligence any more than burying beetles, but they do possess a system of physical mechanisms (the proximate explanation) that equips them with the bodies and behaviors that help them to survive and reproduce (the ultimate explanation).

In dung beetles, a critical period occurs when the larva transforms into an adult (the pupal stage). At this point the presence or absence of the sex chromosome determines what genes will be turned on or off to produce an adult male or female. If it's a male, then body size is assessed and compared to a threshold to determine if a horn will be produced. Other decisions are made as well. The larval body is like a single pot of money that needs to be allocated to adult body parts: how much should go to the wings, eyes, legs, antennae, ovaries, or testes? Somehow, the hormonal interactions inside the beetle function like a calculator, receiving information (such as body size) and giving sensible answers (the right body parts and proportions).

Horns are clearly adaptive for the individuals that have them (large males), but we have yet to explain the *diversity* of horns. Why are they located on the snout, forehead, or shield behind the head in different dung beetle species? It turns out that the answer to this question is based not on a diversity of functions (all horns are used for basically the same purpose) but on a quirk of beetle development. Competition occurs not only between male beetles battling with their horns but between adjacent body parts within male beetles during adult development. If the horn is located on the snout, it can become bigger only at the expense of the antennae becoming smaller. If the horn is located on the forehead, it can become bigger only at the expense of the eyes becoming smaller. If the horn is located on the shield, it can become bigger only at the expense of the wings becoming smaller. These trade-offs might seem puzzling. Shouldn't it be possible to design a system that doesn't require competition among adjacent structures? Perhaps, but that's not the system that evolved in dung beetles.

Internal competition among adjacent body parts begins to explain why horns might be located on different parts of the body in different species. For example, among the thousands of species in the genus *Onthophagus*, some are diurnal and others are nocturnal. Nocturnal species need bigger eyes to see in the dark, so their horns should be located on their snouts or shields but not on their foreheads. Sure enough, that is what Doug discovered when he compared diurnal and nocturnal species within this single genus. Similarly, species that must fly long distances should not have horns on their shields, and species that rely especially on their antennae to sense chemicals should not have horns on their snouts. Doug is in the process of testing these hypotheses by comparing the thousands of species in this single genus in the appropriate ways.

Doug's dung beetles and Belyaev's foxes tell us that natural selection thinking and mechanistic thinking—ultimate and proximate—must take place in combination to fully understand the evolutionary process. The living clay of heritable variation is not infinitely malleable and its properties can only be discovered by hard work. No one could have known beforehand that foxes selected purely for tameness would become like dogs in other respects or that adjacent body parts compete with each other during beetle development. The only way to obtain this knowledge is to roll up our sleeves and get to work using scientific methods and a good theory that asks the right questions.

Doug and I are both evolutionists, but we practice our craft in very different ways. I use evolution to study the length and breadth of creation, spending only a few years on any particular organism or subject. He is basing his entire career on a single subject (horns) and a single esoteric group of insects (dung beetles), which might seem incredibly narrow. But Doug's work is narrow the way a laser beam is narrow.

He uses his detailed knowledge to ask and answer basic questions that I can't begin to approach with my relatively superficial knowledge. Both approaches deserve to coexist for their separate insights.

Doug and I are both basic scientists, which means that our primary goal is to advance knowledge for its own sake. It might seem that basic and practical science should be fully compatible. If you can study evolution and development while breeding a better fox, why not? Alas, such happy combinations are only sometimes available. Just as there are trade-offs between adjacent body parts of a dung beetle, there are trade-offs between basic and practical science. The fox experiment could have been performed much more quickly and inexpensively using wild mice, which happen to have no economic significance. Emlen chose his dung beetles as the very best system for advancing basic knowledge about evolution and development. There is no comparable system that is economically significant. Modern life rests upon a foundation of basic knowledge, and strengthening the foundation does not need to be justified in terms of a more narrow practical goal.

The metaphor of basic science as a foundation can be taken further. Building a foundation requires a certain amount of stability and wealth. There's no point if you won't be here tomorrow or if you are worried about your next meal. Russian society lacks the stability and wealth to invest much in basic knowledge. In fact, the situation after the collapse of Communism is worse than ever, as Belyaev's successors describe in a 1999 article that amounts to a cry for help in their heroic effort to continue the fox experiment:

> For the first time in 40 years, the future of our domestication experiment is in doubt, jeopardized by the continuing crisis of the Russian economy. In 1996 the population of our breeding herd stood at 700. Last year, with no funds to feed the foxes or to pay the salaries of our staff, we had to cut the number to 100. Earlier we were able to cover most of our expenses by selling the pelts of the foxes culled from the breeding herd. Now

that source of revenue has all but dried up, leaving us increasingly dependent on outside funding at a time when shrinking budgets and changes in the grant-rewarding system in Russia are making long-term experiments such as ours harder and harder to sustain. Like many other enterprises in our country, we are becoming more entrepreneurial.

It is fortunate that at least some nations on this earth have the stability, wealth, and wisdom to invest in the advancement of basic knowledge for its own sake.

8

Dancing with Ghosts

IMAGINE A SURREAL DREAM that begins with a ballroom of dancers in elegant attire. Suddenly, one member of each pair vanishes, but their partners dance on as if nothing has changed. Their arms remain outstretched and they continue to circle as if they were dancing with ghosts. Then a bottomless pit appears in the middle of the dance floor. You watch spellbound as the solitary dancers approach the edge of the pit, like actors approaching the edge of a stage. Alas, the dancers are as heedless of the pit as they are of the disappearance of their partners. There is nothing you can do as they plummet one by one out of sight.

This dark fantasy becomes reality every time a species encounters a new environment. I have already shown that creatures such as burying beetles and dung beetles do not have humanlike intelligence. Their wisdom—their dance—is the product of birth and death processes sculpting the living clay of heritable variation over many generations. If you change the environment, nothing mental happens to cause them to change their behavior. The successful strategies simply become unsuccessful and fade slowly away as the living sculpture acquires a new form.

The fact that time is required for adaptations to evolve by natural selection forces me to amend my description in

Chapter 3 of the environment as a quick guide to the properties of organisms. That was too simple. The organisms must
reside in the environment for a sufficient number of generations for the organism-environment relationship to become
established.

How many generations are sufficient? Darwin imagined
evolution as a glacially slow process, operating on tiny differences in survival and reproduction. Jonathan Weiner's marvelous book *The Beak of the Finch* shows how much that
conception has changed based on modern research. The
forces of nature can blow at hurricane strength, and tiny differences between individuals, such as a fraction of a millime-
ter in a finch's beak, can make the difference between life and
death. In some species of zooplankton (minute creatures that
float in the open water) a quarter of the population is consumed by predators *every day*. A single strain of bacteria
placed in nutrient broth will mutate into a variety of forms
that occupy different niches, such as the liquid, the sides of
the container and the surface film, in only a few days (encom-
passing many bacterial generations). In other words, the same
kind of diversification that takes place on remote islands such
as Hawaii and the Galápagos also takes place at the microbial
level in a cup of soup if you absentmindedly forget about it
and come back a week later to find that it has become
"yucky." When you get a bladder infection, it's not just because some *E. coli* found their way to your bladder. It's because genetic evolution took place within your body, creating
a new strain of *E. coli* capable of sticking to your bladder wall.
Natural selection is taking place in and all around us, as eloquently recounted in *The Beak of the Finch*.

Still, natural selection requires *some* time and when a
species encounters a new environment it starts dancing with
ghosts. Since time immemorial, baby sea turtles have emerged
from their nests on beaches at night and made their way
toward the sea. They evolved to rely upon the light reflecting
from the surface of the water, which provided a reliable cue
until the construction of beach houses. Now lights from the
houses cast an even greater glow than moonlight reflecting from the sea, causing the turtles to head in the wrong

direction, toward their deaths, exactly like the dancers falling into the pit of my surreal ballroom.

I could provide dozens of other examples, many caused by human environmental change. Pronghorn antelope flee with amazing speed and endurance from predators that no longer populate the American plains. Oak trees time their acorn crops in response to passenger pigeons that no longer darken the skies. These species will continue dancing with ghosts until they go extinct or until natural selection teaches them new dance steps.

The idea of a new environment requires clarification. The wood frog (*Rana sylvatica*) lays its eggs in small pools of water during the spring. A given pool might or might not have predators. For a tadpole that hatches in a pool with predators, a pool without predators would be a new environment, but the *population* of tadpoles lives in both environments on a regular basis. What adaptations are likely to evolve in this situation? Ideally, a tadpole should be able to first assess its local environment and then display the appropriate adaptation. We have already seen an example of this kind of flexibility in male dung beetles, who assess their size and then display horns only if they are sufficiently large.

I hope that you can see by now that a prediction such as this one is not destined to be correct. It is just a reasonable guess that requires work to confirm or reject. It turns out that many species are very good at assessing their environment and displaying the appropriate adaptations. Wood frogs have a plan A and plan B for the presence and absence of predators, just as dung beetle males do for their own body size. The presence of predators is sensed chemically, and each plan involves a coordinated suite of behaviors (such as movement), anatomical traits (such as the size of the tail), and life history traits (such as when to emerge from the pond). The environment has a huge effect on the organism, but not in the way that we usually associate with learning. Instead, a very specific feature of the environment (the presence or absence of a certain chemical) is used as a switch to activate genetically determined strategies.

As another example, imagine raising a certain species of

minnow from the egg stage in a number of aquaria under carefully controlled conditions. When they are six months old, take a plastic model of a pike (a minnow predator) on the end of a stick and move it slowly through the water for one minute in half of the aquaria. Then do nothing for eighteen more months. That incredibly brief experience has a profound and lasting effect on minnow behavior, causing them to be wary of predators for the rest of their lives. This is not learning as we typically think of it. It is more like an elaborate war plan that is set in motion by a single phone call.

Minnows and wood frogs have environmentally triggered "war plans" to protect themselves against predators because they experienced predators *some of the time* in their natural environment for many generations. How about species that *never* experience predators? These species exist, especially on remote islands. When the first sailors set foot on the Galápagos, the birds treated them not as predators but as trees. They had no war plan for predators because predators had been completely absent from their environment for many, many generations. Lacking humanlike intelligence, they never learned to avoid the sailors, but just allowed themselves to be plucked from their perches and tossed into the cooking pot in another fiendish version of my surreal dream.

What do I mean by humanlike intelligence, and how does it save us from dancing with ghosts? When we are placed in a brand-new situation, we have at least some capacity to realize that we have a problem and work toward a novel solution. A fast mental process takes place that accomplishes roughly the same thing as the slow generational process of natural selection. I will have more to say about humanlike intelligence later, but now I want to emphasize its limited nature. It's a wonderful thing that works some of the time, but if you think that it entirely solves the problem of dancing with ghosts, you are sadly mistaken. Before we had humanlike intelligence, we were mammals and primates with an arsenal of war plans that evolved by natural selection. Our humanlike intelligence was added to the arsenal. It did not replace the other war plans, nor would we necessarily want it to. Some of the war plans aren't even mental. In fact, the whole *concept* of "mental" is dissolving in front of us. A brain is a certain physical system that acts like a calculator, receiving information and giving sensible answers. We have already seen that larval beetles have hormonal calculators that do the same thing. Even bacteria have molecular calculators that take in information and give out sensible answers. We need to stretch the concept of mental to include all types of calculators, regardless of their material composition. When we do this, it becomes clear that we are dancing with ghosts despite our vaunted human intelligence.

Our eating habits provide a compelling example of dancing with ghosts. Our lust for fat, sugar, and salt makes great sense in an environment where these substances were in perennial short supply, but putting a fast-food restaurant on every corner is like lighting up the inland sky for baby sea turtles. We rush to consume, but it is a cruel joke and we end up killing ourselves. We know there is a problem, but that doesn't mean that we can solve it by a simple act of willpower using our wonderful intelligence. Our so-called rational mind simply doesn't have that much control over the rest of our mind and our body.

This much is clear from common experience, but a closer look reveals that we are dancing with *different* ghosts. As far as we know, based on current information, our ancestors left Africa (or stayed, if you are African) approximately seventy thousand years ago and spread over the globe, arriving in Australia approximately sixty thousand years ago and the Americas approximately thirty thousand years ago. We can thank our humanlike intelligence for this expansion, because it required the solution to new problems at a grand scale. Wherever we went, we figured out how to extract food from our environment until we were eating everything from seeds to whales. Then we figured out how to produce our own food, not once but numerous times in various regions of the planet, as recounted by Jared Diamond in *Guns, Germs, and Steel*.

In each separate human population, the slow wisdom of natural selection followed where the fast wisdom of human intelligence was leading. In cultures that tended livestock, milk became an adult resource for the first time in mammalian history. At first it was hard to digest because the mammalian body is adapted to shut down the infant digestive system (plan A) and start up the adult digestive system (plan B) at weaning. A short-term solution was to have microbes do the digesting by fermenting the milk, but then genetic mutations arose that enabled adult humans to directly digest milk and these provided a sufficient advantage to become common in populations that tended livestock. Other human populations started to become genetically adapted to their particular diets in the same way. It is important to stress that all of the genetic mutations were occurring in all of the populations but only increased in frequency when they provided an advantage. Thousands of years provides enough time for this kind of genetic diversification. Slowly, each separate human population started to perform the right dance steps for its environment, only to have the surreal dream repeat itself with the massive migrations and environmental changes of modern times. In the 1950s, American foreign aid programs shipped dry milk powder all over the world, producing massive flatulence in regions where people are not genetically adapted to digest milk as adults. No wonder they hate us! Today, the question of whether you weigh 160 or 300 pounds in the same fast-food environment depends in part on where your ancestors came from.

It also depends in part on what you weighed when you were born. Just as dung beetle grubs have a hormonal calculator that assesses body size, it appears that mammalian species make a similar assessment early in life, before they are even born. If they are small, they take it as a signal that conditions are bad and that they should hoard every calorie as an adult. If they are large, they take it as a signal that food is plentiful and that they can afford to be less efficient as an adult. Of course they are not thinking at all because they are still embryos, but some physical system is functioning as a calculator to implement metabolic plan A or metabolic plan B based upon the environmental signal (body size). Once the decision is made, there's no going back any more than a male dung beetle can reconsider his horn.

This kind of adaptive flexibility evolved because fetal body size was a reliable indicator of the adult food environment for many thousands of generations, even before our appearance as a species, since other mammals such as rats have the same adaptation. We can play a cruel joke on a rat by feeding its mother a restricted diet during pregnancy and then switching the environment by giving the offspring as much food as it wants. The rat becomes obese because its metabolism is designed to squeeze every calorie and now there are so many to squeeze. In just the same way, if you are obese, it is likely that you were underweight as a newborn. You are dancing with ghosts in a surreal dream.

The fact that natural selection takes time and that adaptations are sometimes mismatched with their current environment makes the study of evolution more difficult but also more urgent. We are dancing with ghosts in more ways than our eating habits and our environmental impacts are causing other species to dance with ghosts at an ever-increasing pace. We can't stand by helplessly and watch the tragedy unfold as in the dream. We need to understand how adaptations function and attempt to intervene when they malfunction in our current lives.

What Is the Function of a Can Opener?
How Do You Know?

IN CHAPTER 2, I said that religion, government, and science all face the same problem; individuals and groups are driven to benefit themselves at the expense of other individuals and groups. In science this takes the form of factual claims with hidden self-serving or group-serving messages. You should be on the lookout for suspicious factual claims such as these:

Factual claim	Hidden message
That's sick	Don't do it
That's unnatural	Don't do it
That's immature	Don't do it
That's too hard	Don't do it

Mind you, there are things that are genuinely sick (a person with malaria), unnatural (a bear riding a bicycle), immature (the speech of a toddler), and too hard (moving a piano upstairs by yourself). More often than not, however, these factual claims are incorrect as facts and function to influence behavior, like the messages that are supposedly flashed in movies that say, "Eat popcorn. Eat popcorn."

You might think that only simple uneducated folks employ such crude tactics. Not true. Some of the most important intellectual movements and scientific theories are derived from this source, such as Kohlberg's theory of moral development (we pass through a number of stages of morality as we mature) and Rousseau's idea of the noble savage (natural man is good and corrupted by civilization). Anxiety about evolution is due largely to the visceral assumption that if something is natural (because it evolved by natural selection), then we *can* do it (it will be permissible). To make matters worse, these self- and group-serving tactics can be completely subconscious. I could be guilty of them, for example.

The idea that something is too hard deserves special scrutiny. I have already described science as a roll-up-your-sleeves activity. You return home at the end of the day with sweat on your brow, needing to wash off the odor of carrion beetles, dung beetles, or silver foxes. Lots of things are hard—gardening, basketball, going to college, fighting a fire—but something is *too* hard only when it fails to provide a sufficient benefit. When creationists and ID proponents say that studying evolution is too hard, they mean that a lot of work has produced no conclusive answers.

Creationists aren't the only folks who claim that studying evolution is too hard. Others make the same claim, often under the rallying cry "Just-so story!" Just as Rudyard Kipling spun fanciful tales such as "How the Giraffe Got Its Long Neck," natural selection thinking is portrayed as a form of idle speculation that can seldom be established as scientific fact. Even some famous evolutionists have made this claim, such as Stephen Jay Gould, who felt that some of his fellow evolutionists emphasized adaptation too much compared to such things as development that I discussed in Chapter 7.

With due respect to Steve and others, I think that "Just-so story!" is often used as a code word for "Don't do it!"—a visceral defense reaction by those—including evolutionists—who are nervous about some of the implications of natural selection thinking. As a factual claim it is not only wrong but backward. Establishing whether something is an adaptation is usually *easier* than figuring out such things as mechanisms of development. Moreover, it is pointless to compare their relative difficulty because ultimate and proximate explanations are complementary and can never substitute for each other, as I have already shown.

In my class, I demonstrate these points by passing around a can opener. Everyone knows that the purpose of a can opener is to open cans, but what if someone demanded proof? How do we know that it isn't a mineral dug from the earth, a manufactured object with a different function, or a manufactured object with no practical function, such as a piece of art? The typical response is to say that the object consists of a number of parts that interact in exactly the right way to open cans: a sharp wheel that pierces the metal, a toothed wheel to grab the rim of the can, a lever to apply the pressure that pierces the can, a handle to turn the toothed wheel along the rim, and so on. No mineral is so functionally integrated, it is incapable of performing any other task remotely as well, and I challenge you to find anything in an art museum or a jewelry store that performs an intricate function as well as a can opener opens cans. Only a joker would press the issue further.

Other features of the can opener are more arguable. How about the color or specific shape of the handle? It is possible that these have a function, such as fitting nicely into the palm of the hand, being conspicuous in a drawer crowded with other objects, or attracting the attention of the consumer in a crowded store display, but they might just as easily be arbitrary. In addition, even if the color of a can opener helps *you* to find it in your crowded kitchen drawer, it wasn't necessarily designed with that particular function in mind. How about the hook-shaped notch that exists in many but not all can openers? It's mighty handy for opening bottles, but is that its designed purpose? A bottle opener doesn't have the interacting parts of a can opener, which makes it more difficult to argue with a skeptic about its purpose. Most of these controversial issues about can openers can be settled with a bit of work, but they aren't as self-evident as the function of the major interacting parts.

After discussing can openers with my students, I show them a less common kitchen gadget. It has a handle connected to an oval-shaped metal rim divided into sections. It is obviously utilitarian, but its specific purpose is obscure. Is it a whisk? An egg slicer? Most students don't guess the correct answer, which is an avocado slicer. Once I tell them, however, they immediately regard it as obvious in retrospect. This is because the object is much better suited for slicing avocados than the previously imagined functions, a fact that could not be appreciated until the correct function was discovered. You might have had the same experience browsing in antique stores, looking at objects whose functions were obvious to our grandparents but not to us.

These ways of thinking about function, which my students employ without any formal training, are perfectly valid. An object with a function must have certain design features to perform its function, which are often sufficiently complex and integrated that they cannot be explained any other way. We use our ability to infer purpose all the time and would be helpless without it. The human brain might even be genetically adapted to infer purpose, just as chickadee brains are genetically adapted to remember the location of thousands of seeds over the winter.

Nothing changes when we think about natural objects instead of man-made objects. The function of the heart and circulatory system was obscure until the British physician William Harvey made it self-evident in the early 1600s. Beetles burying mice and battling with horns are obviously behaving in a purposeful fashion. Nature has always and correctly been regarded as a cornucopia of function. Darwin provided a new way of explaining *how* organisms become well designed, but *that* they are well designed (for the most part) was never in doubt, any more than the function of a can opener. Against this background, it is bizarre to claim that natural selection thinking is irredeemably speculative.

By contrast, consider the challenge of figuring out the ancestry of species in the great branching tree of life (phylogeny) that was dear to the heart of Steve Gould. This is a difficult job because the fossil record is incomplete and living species can be similar *either* because they share a common ancestor *or* because they have converged on the same traits despite being distantly related. For example, dung beetle species with horns on their snouts might be similar because they share a common ancestor, or because they independently adapted to a nocturnal way of life. Advances in molecular biology have revolutionized our ability to resolve questions such as these. Sometimes the phylogenies based on older methods are confirmed by the new methods, but often they prove to be spectacularly wrong. In the case of dung beetles, horn location says very little about common descent because it easily changes over evolutionary time within any given lineage—a fact that required the hard work of scientists such as Doug Emlen to establish.

Even though it is hard work to understand phylogenies and developmental mechanisms, it is not *too* hard because the information is important, however difficult to obtain. Good information about our own ancestry is incredibly difficult to obtain, but each success is celebrated as a breakthrough, and converging lines of evidence yield a surprisingly detailed account with enough hard work. When I see documentaries of anthropologists excavating fossils with dental tools and toothbrushes in the baking Ethiopian sun, guarded against bandits by local tribesmen with rifles, I know that it would be too hard for me, but I am thankful it is not so for them.

Evolutionary science is like any other kind of science. A theory generates alternative hypotheses, which are evaluated with scientific methods. Every cycle of hypothesis formation and testing is like turning the crank of a wheel that generates factual knowledge. Evolutionary theory is exceptionally broad in its explanatory scope, thanks largely to natural selection thinking. Hypotheses generated by natural selection thinking are not more difficult and are often simpler to test than hypotheses based on proximate mechanisms and historical processes. Comparing their difficulty is pointless, however, because they are complementary sources of information and the crank must be turned for all of them. To give you an idea of how fast the natural selection crank is turning, I just typed the phrase "guppy natural selection" into Google Scholar and received over 3,000 hits—for just one species of fish! Multiply that by thousands to include research on other species, and it is clear that knowledge is advancing by the minute. Properly understood, "Just-so story" is just another phase for "untested hypothesis" and should be treated as a rallying cry for another turn of the crank.

10

Your Apprentice License

CONGRATULATIONS! I HAVE FINISHED conveying the barest essentials of evolutionary theory, as I said I would do in Chapter 1. If you have accompanied me this far, you can regard yourself as an apprentice evolutionist. For the rest of the book we will use the essentials to explore the length and breadth of creation, from the origin of life to human value systems, developing our competence along the way.

Here is a review of the basic elements so they can remain fresh in your mind. First and foremost is the principle of natural selection, which I described as a third way of thinking, different from both theology and materialism. Natural selection is like an artist molding the living clay of heritable variation. Without the principle of natural selection, our prospects for understanding the world around us are as hopeless as understanding a sculpture without any concept of an artist. Moreover, unlike the inscrutable intentions of the gods and other intelligent agents imagined throughout human history, the shaping influence of natural selection is often so transparent that even a rank amateur can make intelligent guesses about the properties of organisms, as I showed for the subject of infanticide in Chapter 3.

The second major point to grasp is that evolutionary adaptations do not always correspond to what we regard as good or useful in the everyday sense of the word. Indeed, they can be the epitome of shortsighted selfishness that destroys the quality of life for everyone in the long run. This is a bitter pill to swallow and undoubtedly accounts for the reluctance of many people to accept evolutionary theory, no matter how well it has been factually established. Fortunately, the bitter pill turns out to be good medicine. Evolutionary adaptations *include* what we regard as good and useful, even though they are not *confined* to them, as I showed in Chapter 5. This simple but profound insight leads to a form of practical optimism, enabling us to enhance the quality of life by providing appropriate conditions, as I will show repeatedly in future chapters. The practical benefits of a theory that explains the good, the bad, the beautiful, and the ugly far outweigh the comforting but false portrayal of evolution as somehow benign in its entirety.

The third major point to grasp is that the living clay of heritable variation is far more complicated than real clay. Your genes reside in you because they had a net positive effect averaged across all of the individual organisms and environments they have inhabited over thousands of generations. Whether they have a net positive effect on *you* is an open question, as I showed in Chapter 6. Development is a complicated process that connects traits as seemingly different as friendly behavior and a curly tail, as I showed in Chapter 7. These and other complications account for the existence of traits that could never be explained on the basis of natural selection thinking alone, underscoring the need to understand the physical basis of life (proximate explanation) in addition to its functional basis (ultimate explanation).

The fourth major point to grasp is that natural selection takes time and that organisms are frequently out of kilter with their present environment. When this happens, they will be dancing with the ghosts of their past environments, to their own detriment in their current environment, until the slow hand of natural selection teaches them the right moves for dancing with their current partners. Only species that possess a fast process of adapting to their environments can avoid this fate, but even the vaunted mental processes associated with human intelligence do not entirely prevent us from dancing with ghosts, as I showed in Chapter 8.

The third and fourth points required me to back away from my original portrayal of natural selection thinking as being like a heat-seeking missile that unerringly finds its target. For a better metaphor, suppose that you are the head of a wilderness rescue unit who receives an urgent call about a missing child. Your first task is to use all of the information at your disposal to determine where to begin your search. Obviously, you are not going to parachute into the forest and land directly on top of the child, or else the child would not be lost. Nevertheless, your task will be far easier if you can land within ten miles than within five hundred miles, so your first estimate is the most important. After you begin your search, you continue to use all of the information at your disposal, adding what you have learned during the search.

Scientific inquiry is much like this. We use a theory and available factual information to make the best possible first guess. Then we continue to use the theory along with accumulating information to refine our search until we find what we are looking for. Natural selection thinking always plays a role in this inquiry to the extent that the object of our search has been shaped by natural selection. It never plays an exclusive role, to the extent that the third and fourth points outlined above come into play. It sometimes plays an especially important role at the *beginning* of an inquiry because the shaping influence of natural selection is often easier to predict than the details of genetics, development, and phylogeny. This is only a practical consideration, however, and there is nothing else that privileges natural selection thinking compared to other sources of information.

The rescue search metaphor highlights the fifth and last major point that must be grasped to become an apprentice evolutionist: no theory leads directly to the facts. There is always a repeating process of hypothesis formation and testing that I described in Chapter 9 as turning a crank. If you can't turn the crank, then scientific inquiry comes to a halt. Science is often portrayed as an exalted and difficult activity accessible only to an elite caste of intelligent and highly trained individuals. I have made every effort to portray it as a down-to-earth activity, like farming, brick making, and house building. As Edison said about inventing, it requires more perspiration than inspiration, and even the inspiration is often of the "Why didn't I think of that before?" variety that can occur to anyone, usually after a lot of perspiring. If you have accompanied me this far, then *you* have the makings of a scientist, who with a little clear thinking and a lot of hard work can help to create something both personally gratifying and larger and more durable than yourself.

11

Welcome Home, Prodigal Son

THE PRODIGAL SON LEFT home with an inheritance that he foolishly squandered on a profligate life. Destitute and ashamed, he returned to his father's house asking only to be treated as a servant, since he clearly deserved no more. To his surprise and gratitude, he was received with love and forgiveness as one reborn to a new and more sustaining way of life.

Our conception of ourselves as set apart from the rest of nature is a bit like the prodigal son leaving home with an enormous inheritance. The repeated collapse of past civilizations and uncertain fate of our own is like squandering our inheritance on a profligate life. Before we become truly destitute and ashamed, perhaps it is time to return home to a conception of ourselves as thoroughly a part of nature. Perhaps this can lead to a more sustaining way of life in the future than in the past.

In the following chapters we will explore what it means to be 100 percent a product of evolution. First, however, it is important to cultivate the appropriate state of mind. The prodigal son experienced his change of heart only after being defeated by his former habits. Similarly, religions around the world cultivate an attitude of humility and contrition in preparation for acquiring a new faith. Rehabilitation programs such as Alcoholics Anonymous are only marginally religious but still instill a belief that one's former life was worthless and destructive in preparation for acquiring a new and more sustaining lifestyle. If we are to adopt a new set of beliefs about ourselves, it will help to appreciate how our old beliefs have failed.

First, we must abandon the notion that some special quality was breathed into us by a higher power. This does not require abandoning religious faith—*many* people manage to combine a vibrant religious faith with a fully naturalistic conception of the world—but it does require abandoning certain kinds of religious faith. One of my irreverent friends likes to say that the strongest test of faith in a God that intervenes on your behalf occurs when your car breaks down. If you leave it by the side of the road and pray for it to be fixed, you are a true believer. If you fix it yourself or have it towed to the nearest mechanic, you are tacitly acknowledging that some things have a purely naturalistic explanation. Perhaps God created the laws of physics, but the laws are fully sufficient to explain what's wrong with your car without any other consideration of a higher power. If you have the kind of religious faith that causes you to pray for your car by the side of the road, then you are handicapping yourself, your loved ones, and probably the rest of society. If you have the kind of religious faith that permits you to fix your car, then you are doubly fortunate. You can enjoy the advantages of knowledge about the physical world and the considerable benefits of religious faith that I will document in subsequent chapters.

What goes for knowledge of the physical world also goes for knowledge about ourselves. If something is wrong with your body, your mind, or society, it has a naturalistic explanation, just like the problem with your car. Believing that we have special god-given abilities is like praying for your car on the side of the road. Actually, this comparison is not quite correct, because the welfare of our bodies, minds, and societies depends very much on what we believe. Believing that we have special god-given abilities might arguably make us healthier and happier in every way. That would not make it factually correct, however. To be precise, the statement

"Believing that we have god-given abilities is good for us" could be factually correct, while the statement "We have god-given abilities" is in all likelihood incorrect. I make this statement with confidence, not because I am hostile to religion, but because supernatural explanations of ourselves have proven their inadequacy many times over, along with supernatural explanations of the physical world and the rest of nature. The next time that you visit a doctor, you should hope that she is enlightened enough to appreciate the importance of belief for physical and mental health, but you should also be glad that she doesn't resort to supernatural explanations any more than your car mechanic.

Abandoning supernatural explanations is only the first step in our multistep road to recovery. The world is full of people who have already abandoned supernatural explanations, who fully accept the fact of evolution and human origins, and yet haven't a clue about what evolution can tell us about our bodies, minds, and societies in any detail. Your aforementioned doctor is probably among them, as I will show in the very next chapter. The secular belief that we stand apart from the rest of nature takes a variety of forms, but most emphasize open-ended abilities such as learning, language, culture, and rational thought. These capacities supposedly enable us to play by different rules than other species and do not require a detailed knowledge of evolution to understand, even though they presumably arose by a process of genetic evolution. A common claim is that "biology" sets broad limits to our behavior, such as eating and procreation, but that "culture" determines what we do within the broad limits, such as making art rather than babies. It is true but boring to point out that we like to eat and have sex; far more interesting is our rich cultural diversity, about which evolution has nothing to say. Above all, we can choose our own destiny because our behavior is not genetically determined, unlike all other species. Add the appropriate mood music, and humanity becomes like Captain Kirk and the starship *Enterprise*, confidently going where no one has gone before.

Hubris, all hubris! In the first place, it places too much

emphasis on our uniquely human attributes. In the second place, it fails to appreciate how much evolution is required to understand our uniquely human attributes.

Our unique attributes evolved over a period of roughly 6 million years. They represent modifications of great ape attributes that are roughly 10 million years old, primate attributes that are roughly 55 million years old, mammalian attributes that are roughly 245 million years old, vertebrate attributes that are roughly 600 million years old, and attributes of nucleated cells that are roughly 1,500 million years old. If you think it is unnecessary to go that far back in the tree of life to understand our own attributes, consider the humbling fact that we share with nematodes (tiny wormlike creatures) the same gene that controls appetite. At most, our unique attributes are like an addition onto a vast multiroom mansion. It is sheer hubris to think that we can ignore all but the newest room.

I have already provided one example of an adaptation that we share with other mammalian species: the assessment of resources during the fetal stage of development that determines the metabolic strategy for the rest of the òrganism's lifetime. This single adaptation has a combination of features that boggles the conventional mind. We usually don't think of a fetus making an informed decision. We associate decision making with conscious thought or at least with brains, whereas this decision is presumably unconscious and involves a physical system that qualifies as a "calculator" but is not necessarily restricted to the brain. We associate environmental effects with learning, whereas this one is more like a phone call that triggers an elaborately coordinated and previously prepared "war plan." If this is how we are constructed as a species, along with monkeys, pigs, and rats, *we need to know about it* for the sake of our children, not just esoteric scientific understanding.

Now multiply this example by dozens or hundreds of similar examples, influencing all aspects of our bodies, minds, and societies, and you will begin to appreciate the need to think of ourselves as a product of evolution, just like any other species. A good recent book on this topic is *Strangers to Ourselves:*

Discovering the Adaptive Unconscious, by social psychologist Timothy Wilson, who shows how many of our decisions are driven by unconscious algorithms similar to fetuses "deciding" their metabolic strategy.

After we appreciate that we live in a multiroom mansion built by evolution, we can turn our attention to the newest room that distinguishes us from other species. Over the centuries, we have regarded ourselves as uniquely intelligent, moral, flexible, and capable of aesthetic appreciation. Most of these are self-congratulatory and suspect as factual claims. I have already shown that other species far surpass our intelligence for specific tasks and that traits associated with goodness can evolve in any species, given the right environmental conditions. I will expand upon these themes in later chapters and also reveal the deep evolutionary roots of aesthetic appreciation. Nevertheless, at the end of the day there is no denying our uniqueness as a species, especially when it comes to our behavioral flexibility and ability to construct our own social environments. Many professors and intellectuals call themselves "social constructivists" and glory in cultural diversity. Their problem is not that they are wrong but that they see their position as non-evolutionary. They need to become *evolutionary* social constructivists, as I will now show in a roundabout way.

Consider the mammalian immune system. It is a fabulous adaptation that evolved to protect us from the tiny predators that we call disease organisms. Big predators such as lions and tigers loom large in our imagination, but the tiny predators are more deadly and are still with us, invading our bodies with every breath and bite we take. When we die, our immune system stops functioning and we are immediately consumed by microbes, like billions and billions of hyenas fighting over a vast elephant carcass. Somehow our immune system keeps these predators at bay, but how? One possibility is a large number of pre-evolved "war plans," similar to the allocation strategies of dung beetles, predator defense strategies of wood frogs and minnows, and metabolic strategies of fetuses that I have already described. Microbe X invades our body, is detected chemically, and immediately countered with war plan X from the arsenal of the immune system. This is *partially* how the immune system works, but it is not and cannot be the whole story. Not only are there too many different types of microbes but each type is evolving at a very fast rate, similar to the microbes in your cup of soup that I described in Chapter 8. The only way to combat such a diversity of fast-changing foes is to fight fire with fire, evolution with evolution. The heart of the immune system, as you probably already know, is the production of antibodies at random and the selection of those that successfully bind to the particular disease organisms in your body. *The immune system is a fast-paced process of antibody evolution, created by the slow-paced process of genetic evolution.*

Unlike the other genetic adaptations that we have considered so far, the immune system solves the problem of dancing with ghosts. Suppose that a totally new type of microbe arrives from Mars and invades your body. The random production and selection of successful antibodies might still be an effective defense, regardless of whether the microbe came from earth, Mars, or the next galaxy. The fact that the immune system is an evolutionary process in its own right means that it can evolve adaptive solutions on its own time scale rather than on the time scale of genetic evolution.

Now suppose that someone were to argue that we can understand the immune system without bothering to think about evolution. This would be deeply wrong for two reasons. First, it would ignore the fact that the immune system at its core is a fast-paced evolutionary process. Second, it would ignore the fact that the immune system arrives at adaptive solutions only because of a vast and complicated architecture that evolved by genetic evolution. Understanding the immune system requires detailed knowledge of evolution operating at two different time scales.

I hope you can see how my roundabout route leads back to the subject of human uniqueness and social constructivism. The immune system is not the only fast-paced evolutionary process built by genetic evolution. Gerald Edelman, who received a Nobel Prize for his work on the immune system, went on to study the human brain as an evolutionary process in its own right, which he calls "neural Darwinism." Most of the open-ended processes that we associate with human uniqueness, from flexible brain development to symbolic thought and cultural diversity, reflect fast-paced evolutionary processes that take place within an architecture created by genetic evolution. We have not escaped evolution. We experience evolution at warp speed. The starship *Evolution* is not like the starship *Enterprise*, however. Unless we understand how it works, it will take us to places that we don't want to go.

Like the prodigal son and programs such as Alcoholics Anonymous, I have tried to cultivate a sense of humility about old ways of thought in preparation for encountering new ways of thought. Let's recount the steps that we have taken on our multistep path to recovery. First we had to abandon the notion that we have special qualities breathed into us by a creator. Then we had to acknowledge that we live in a mansion with many rooms, most shared with other species. Finally we had to acknowledge that our special room requires the same detailed understanding of evolution operating at multiple time scales that is required to understand the immune system. These conclusions follow from evolutionary theory at such a basic level that they are very unlikely to be wrong, but they will require seismic adjustments in conceptions of human nature that are centuries old. As I said in Chapter 2, even though Darwin's theory is nearing its 150th birthday, so much remains for the future. Now let's explore what it means to think of ourselves as 100 percent a product of evolution and how this way of thinking can help us discover a more sustainable way of life.

12

Teaching the Experts

I BEGAN THIS BOOK with a list of tall claims about evolution: that it can become uncontroversial, that the basic principles are easy to learn, that everyone should want to learn them, and so on. By now I hope you agree that these claims are tall but not outrageous. They are tall *supportable* claims that follow from evolutionary theory at an elementary level.

One of my tall claims is that an evolutionist such as myself can waltz into a new subject (such as religion) and teach something to the experts whose factual knowledge is far greater than I can ever hope to achieve. If you aren't familiar with natural selection thinking, then you are like Darwin before his moment in the carriage or a person trying to explain a sculpture without any concept of an artist. If you are an expert, then your problem might be even worse because your head is filled with facts on a narrow subject that prevent you from considering other subjects. No one illustrates these points better than Margie Profet, a woman who rocketed to fame in 1993 by becoming one of the youngest recipients of a MacArthur Foundation "genius" award for her evolutionary theory of pregnancy sickness.

Margie wasn't an expert at anything. A daughter of egghead parents (an engineer mother and a physicist father), she majored in political science at Harvard and obtained a second degree in physics at the University of California at Berkeley. Neither degree satisfied her intellectual wanderlust, so she dropped out of academia and became a self-described "bum" to give herself more time to think. She had no formal training in evolution or even biology, but as she described in a 1996 *Scientific American* article, "I knew some people in evolutionary biology, and I would have some conversations with them, and I would read everything, and I just started thinking about things."

One of those things occurred to her during a conversation with some pregnant relatives who were complaining about their morning sickness, which is better named pregnancy sickness because it can occur at any time of day. Like many other pregnant women, they couldn't eat certain foods that had been a part of their normal diet only a few weeks before, including favorites such as coffee, and foods that are supposed to be good for you, such as dark green vegetables. Pregnant women often become nauseous at the mere smell of these foods and sometimes are so violently ill that they seek a doctor for help.

These familiar facts about pregnancy sickness didn't sit well with Margie's newly acquired knowledge of evolution. Taken by itself, pregnancy sickness is just plain bad for the pregnant woman and her baby-to-be. Natural selection weeds out things that are just plain bad. Something must be added to the story to explain why millions and millions of women mysteriously get "sick" every time they get pregnant. Margie's reasoning about pregnancy sickness was similar to Steve Suomi's search for a benefit to counterbalance the cost of the psycho male monkeys described in Chapter 6.

Margie's nagging question had more than one plausible answer. Perhaps pregnant women are especially vulnerable to disease and their sickness is caused by an infectious agent. Perhaps there is a widespread environmental pollutant that causes pregnancy sickness. Perhaps pregnancy sickness is unavoidably connected to hormonal changes that take place during pregnancy, just as a curly tail is connected to friendly behavior. Or perhaps the cost of pregnancy sickness is associated with some larger benefit that can be identified with a bit of natural selection thinking.

It didn't take long for Margie to devise a plausible theory of pregnancy sickness as an adaptation, just as my students (and perhaps you) were able to devise a theory of infanticide. Most species are locked in an evolutionary race with their predators and prey. Antelope and cheetahs evolve to run faster until both are running at lightning speed. Seeds evolve thicker shells and parrots evolve thicker beaks until they are cracking nuts that you would have difficulty opening with a pair of pliers. A similar evolutionary race has created countless species of animals, plants, and microbes that are adapted to protect themselves with toxins but nevertheless are eaten by species that are adapted to cope with the toxins. As an omnivorous species descended from a long line of omnivores, we are so good at coping with the chemical defenses of other species that we actually use them to fight our own parasites and diseases.

Margie reasoned that embryos might not have the ability to withstand the chemical onslaught that is part of the daily adult diet. If not, then pregnancy would require a change in the eating habits of the mother to protect her developing embryo from being poisoned by the toxins in her normal diet. The change would not be conscious, however, but would happen with the same mechanical regularity of a beetle developing horns or a fetus adopting a given metabolic strategy. Pregnancy sickness might be uncomfortable for the mother and deprive the developing embryo of calories, but the consequences of *not* experiencing pregnancy sickness could be much worse.

Margie's theory was appealing, but that didn't make it correct. The best that a theory can do is suggest a number of plausible alternatives, as I have already stressed. Further progress requires turning the crank of the scientific method. Fortunately, very different consequences follow from the various hypotheses. *If* pregnancy sickness is caused by an infectious agent, *then* it should be possible to identify the agent or perhaps cure the sickness with antibiotics. *If* pregnancy sickness is caused by a widespread pollutant, *then* it should be a recent phenomenon and perhaps most common in industrial nations. *If* pregnancy sickness is a by-product of hormonal changes taking place during pregnancy (like the dog's curly tail), *then* it should be possible to discover the hidden connections. *If* pregnancy sickness is an adaptation to protect the developing embryo from toxins, *then* a number of consequences follow. It should be timed to coincide with the most vulnerable period of development, it should be triggered by the foods most likely to harm the embryo, and so on. Each hypothesis leads to such different predictions that it should be straightforward to determine the truth of the matter with enough hard work.

Working from a studio apartment that she shared with squirrels and scrub jays who came to receive handouts of peanuts, Margie immersed herself in the scientific and medical literature. Recall from Chapter 4 that facts are like bricks—durable and easily produced but worthless until assembled into a larger structure. Darwin's great achievement was to provide a theory that organizes existing facts and directs the search for new facts. Margie's challenge was to repeat Darwin's achievement for the specific subject of pregnancy sickness. She had access to hundreds of facts, but they were like bricks piled haphazardly here and there. Could she assemble them into a sturdy structure, and which hypotheses would the structure support?

One by one, the facts seemed to fall in favor of the adaptation hypothesis. As early as 1940, one medical researcher reported that women who experienced severe pregnancy sickness were much *less* likely to miscarry than women with less severe symptoms. Pregnancy sickness occurs mostly during the period that the embryo is developing its major organ systems and is most sensitive to toxins. Spicy and bitter foods are more likely to trigger pregnancy sickness than bland foods. These same foods are implicated in miscarriages and birth defects. There is no known infectious agent or environmental pollutant associated with pregnancy sickness. Pregnancy sickness is not confined to modern industrial nations.

Margie's investigation soon led her beyond the narrow subject of pregnancy sickness to other adaptations that appear designed to protect the developing embryo from food toxins. Not only do pregnant women avoid certain foods, but their bodies work harder to detoxify the food that they eat. Food moves more slowly through the intestines. Blood flow increases to the kidneys. The liver steps up enzyme production. The nose becomes more sensitive to odors. Even the seemingly bizarre habit of eating clay falls into place, since clay has been shown to reduce the absorption of toxic chemicals into the bloodstream and is a primary ingredient in Kaopectate, which is used to treat upset stomachs and nausea. These coordinated changes have all the earmarks of a major physiological "war plan" that evolved over millions of generations, long before our appearance as a species, to solve a recurring problem of survival and reproduction.

Let's pause to savor the irony of the situation. If Margie's theory is correct, then women are biologically adapted to protect their babies during development. They have no conscious knowledge of their ability, however, and only experience the distressing symptoms of food aversion. They visit their doctors, thinking that they are sick. The doctors are learned experts who almost certainly believe in evolution, but the vast majority don't think about evolution in relation to their profession. Their main concern is to find something that will prevent pregnant women from barfing. In the 1950s, they widely prescribed a drug called thalidomide, which caused tragic birth defects in thousands of children worldwide. Even after this particular drug was discontinued, virtually no one questioned the interpretation of pregnancy sickness as a *sickness* that needs to be *cured*. It took an intellectually curious bum to put the big picture together like a jigsaw puzzle, using natural selection thinking as the picture on the cover and the facts generated by experts as the pieces, in her apartment surrounded by her friends the squirrels and scrub jays.

Margie published her results for two audiences. For the experts she wrote a long article in an edited book titled *The Adapted Mind: Evolutionary Psychology and the Generation of Culture* that was published in 1992 and helped to initiate the modern study of our species from an evolutionary perspective. For mothers everywhere she authored a book titled

Protecting Your Baby to Be: Preventing Birth Defects in the First Trimester, published in 1995. Some members of the medical establishment clucked disapprovingly that Margie was dispensing unproven medical advice, but who were they to cast blame? Welcome to the world of incomplete information. It's true that Margie stitched her argument together from past studies that were conducted largely for other purposes, but was that her fault? Why didn't it occur to the experts that pregnancy sickness might represent the downside of an important adaptation? What are they going to do about it in terms of future research? Why should mothers everywhere be prevented from learning about Margie's theory, which, after all, is based on a careful evaluation of past scientific information?

Pregnancy sickness is the first topic that I introduce to my students after teaching them the basic principles of evolution. Rather than using Margie's paper, I use a more recent review article by Samuel M. Flaxman and Paul W. Sherman, published in 2000. Paul is an evolutionist of my vintage who became famous even as a graduate student for his research on a certain species of ground squirrel. He would arrive at his field site every summer with a team of undergraduate students who sat in lifeguard chairs, literally from dawn to dusk, recording everything that the ground squirrels did when they were aboveground. Paul was one of the first to provide evidence for the theory of kin selection, which predicts that animals should be especially nice to their genetic relatives. I still remember that when he entered the job market, he was the first choice of so many universities that everyone else competing for the same jobs had to wait for Paul to make his decision. He ended up at Cornell University and studies a diversity of subjects, just like me. Sam was an undergraduate student at Cornell who worked with Paul to evaluate Margie's theory. In other words, both Sam and Paul were nonexperts on the subject of pregnancy sickness, even though Paul is a master evolutionist.

They say that bricklaying looks simple but is actually a highly skilled task. A master bricklayer is a sight to behold, applying the mortar and positioning the bricks without a single wasted motion. That is how I think of the article by Sam and Paul: a master and his apprentice at work, turning the crank of the scientific method without a single wasted motion. Most of my students agree, which gives me great pleasure. They have been thinking about evolution for only a few weeks, and already they are reading scientific articles—written for the experts—not only with understanding but with pleasure.

In Chapter 3, I said that natural selection thinking can be applied to an infinite number of subjects. Margie's mind was on fire for three subjects, not just one. In addition to pregnancy sickness, she had theories about allergies and menstruation. Her reasoning was the same in each case: taken by themselves, these things are *bad* and would be eliminated by natural selection. What are they connected to that maintains them as a part of human experience? This simple reasoning process is not like a heat-seeking missile that unerringly finds its target, as I have already stressed. It only parachutes you into the general vicinity of an answer, but who in their right mind would avoid using such information when it is so easily available? As far as I know, Margie's theory of menstruation has not fared well, her theory of allergies has fared better, and both have helped to spin the wheel of the scientific method faster. As for Paul Sherman, with the help of another undergraduate student named Jennifer Billing, he gave another command performance on the seemingly arbitrary subject of why we spice our foods. Have you ever wondered why some cuisines are so much spicier than others? Did you know that when the Goths besieged Rome in 408 CE, they demanded as ransom five thousand pounds of gold and three thousand pounds of *pepper*? Just as my students are beginning to read scientific articles with understanding and pleasure at this point in my course, you might enjoy visiting Paul's Web site (just type his name into Google) and downloading some of his articles on spices. They are more interesting than the entertaining fluff in popular magazines, once you are suitably prepared. As one of my students commented, "I'll never think of spices the same way again!"

Medical science is highly sophisticated in its own way, but

it seldom avails itself of natural selection thinking. Most doctors and medical researchers believe in evolution as a matter of course, but their exposure to evolutionary theory in medical school is close to zero and they don't think about it in relation to their profession. As long as this situation continues, intellectual bums like Margie Profet and master evolutionists such as Paul Sherman with his cadre of undergraduate students have much to teach the experts.

13

Murder City

IT WAS JUNE 5, 2005, and I was in the lounge of the upscale Hyatt Regency Hotel in Austin, Texas. I had gathered with 450 of my brethren to attend the annual conference of the Human Behavior and Evolution Society (HBES). By chance, twenty thousand bikers had also converged upon Austin to attend the 2005 Republic of Texas (ROT) biker rally. Monster machines lined the parking lot, and tough-looking ROTers mingled with bookish-looking HBESers at the bar. This might seem like the opening scene of a cataclysmic culture war, but it was nothing of the sort. Everyone was there for a good time, and there wasn't even a hint of tension.

The previous night I had mingled with HBESers, ROTers, and others at one of Austin's greatest attractions, the world's largest urban bat colony. We gathered at one of the bridges crossing the Colorado River as if we were attending an open-air concert. By the time it was getting dark, the grassy area below the bridge and the bridge itself were filled with spectators, and the first Mexican free-tailed bats began to stream out from under the bridge. The stream turned into a torrent that kept on coming until it boggled the mind how so many bats—estimated at 1.5 million—could possibly fit under a single bridge. How wonderful that this outpouring of life has become a cause for celebration rather than extermination.

Although they were invisible to us, hundreds of thousands of baby bats remained under the bridge awaiting the return of their mothers. They hang upside down and are packed so closely together that they form a continuous living carpet of bodies. You might be wondering how a given mother can find her own child among so many. Strictly speaking, it isn't necessary. Mothers could hook up with any baby and the population would perpetuate itself, perhaps even better than if each mother insists on finding her one and only. On the other hand, a little natural selection thinking makes it clear that mothers who preferentially feed their own children will perpetuate their genes better than mothers who feed indiscriminately.

As with every other prediction in this book, this one is merely an educated guess that isn't necessarily correct. Perhaps there is simply no way for mothers to find their own nubbin in the living carpet, no matter how advantageous it would be. As it happens, this question has been asked and answered for the Mexican free-tailed bat. Each baby has a signature squeak that the mother can recognize among hundreds of thousands of other squeaks. The proximate mechanisms that make this possible are even more amazing than the sight of the endless stream of bats that we have gathered to see. The newborn baby must make its unique squeak, that squeak must be imprinted in the mind of the mother before her very first flight away from her child, and the mother must have the auditory equipment to identify and locate that one squeak amidst a din of other squeaks. This ability is superhuman because we were never faced with this particular problem in our evolutionary past. After we marvel at its complexity, we must also acknowledge its dark side. A baby who loses its mother is not fed by anyone else. It is surrounded by bodies but not by love. Its signature squeak grows more strident, then fainter, until the baby bat drops into the Colorado River and becomes fish food.

Night fell and I joined my fellow HBESers streaming back to our hotel. The four-day HBES conference featured hundreds of talks and posters, similar to the example of pregnancy sickness that I described in the last chapter and collectively covering the length and breadth of human experience. Some of the biggest scientific celebrities were there, including Ed Wilson and Steven Pinker, in addition to others who rank as stars among the HBESers, even if less well known to the general public. These included Martin Daly and Margo Wilson, who published the landmark book *Homicide* in 1988. *Homicide* did not achieve the readership of Ed's *Consilience* or Steve's *The Blank Slate*, but in some ways it was even more important because it did some scientific crank turning within the pages of the book, as opposed to discussing lofty generalizations, as important as these might be.

Martin and Margo had the brilliant idea that homicide statistics could be used to test evolutionary hypotheses about human behavior. After all, nearly every death is recorded and investigated, even if not every case is solved. People kill each other for reasons that they feel passionate about, unlike undergraduate students filling out yet another psychological questionnaire. Finally, if evolutionary theory can shed light on why people kill each other, it can have enormous practical value in addition to its purely scientific interest. *Homicide* presents the results of their inquiry. Why do parents kill their offspring? Why do offspring kill their parents? Why do husbands kill their wives and vice versa? Why do strangers kill each other, often over seemingly trivial matters? Why is homicide so often practiced among groups and legitimized as warfare? For each of these questions, Daly and Wilson piece together the evidence like a jigsaw puzzle, using natural selection thinking as the picture on the cover and crime statistics as the pieces, just as Margie Profet did for the subject of pregnancy sickness. I still remember when I received the book by mail, with the single word *Homicide* printed on a black cover as if from the typewriter of a hard-boiled crime reporter. I read it immediately and loved it so much that I gave it to all my family and friends for Christmas. For me it was much better than a crime novel because it was approaching the truth: facts as durable as we can establish, assembled into the most robust explanatory structure that we can construct.

I asked Martin and Margo to have breakfast with me during

the conference to describe how they came to write their wonderful book. They are a husband-and-wife team who have worked together for so long that their conversation seamlessly alternates, one picking up where the other leaves off, with occasional humorous disagreements like well-practiced comedy routines. Their story begins in the 1970s, when they were young assistant professors at the University of California's Riverside campus. They were studying animals then, not humans, and were reading Ed Wilson's newly published *Sociobiology: The New Synthesis* with a group of graduate students.

Ed's book reflected a transformation that had taken place in the study of animal behavior. Even though the seed of natural selection thinking was planted by Darwin and grew during the early twentieth century, thanks to pioneers such as Karl von Frisch, Konrad Lorenz, and Niko Tinbergen (who were awarded a Nobel Prize in 1973), it wasn't until the 1960s that scientists started to routinely ask the simple question "How *would* a well-adapted animal behave in a given situation?" as I asked for the subject of infanticide in Chapter 3. Hundreds of predictions emerged, as if from a crystal ball. Mathematical models gave these predictions an uncanny precision: foraging predators should ignore a given prey type precisely when the time required to process it can be better spent searching for and processing more desirable prey. Individuals of the same species should be inclined to help each other in direct proportion to their coefficient of genetic relatedness. The sex ratio of parasitic wasps should depend on how many females lay their eggs on a single host. Many biologists unaccustomed to natural selection thinking were incredulous. Why should a wasp behave in a way that requires calculus for us to predict? Love them or hate them, the predictions were eminently testable, and the crank of the scientific method began to turn with a satisfying whir. A new picture emerged of animals as far more sophisticated in their behavioral strategies than most scientists previously imagined. *The New Synthesis* was an apt subtitle for *Sociobiology* because, for the first time, all forms of social behavior in all species were being approached from a single theoretical perspective. Ed's final chapter on humans created an uproar, but few people challenged the new synthesis for the rest of the animal world.

Martin and Margo were impressed with the predictive power of evolutionary theory but also prided themselves on being good empiricists whose motto might be "Show me the data." Martin was studying parental care and neglect in mice (similar to my work on burying beetles described in Chapter 4), and the discussions of *Sociobiology* often turned to evidence in other species. Almost as an afterthought, someone asked, "What about humans?" After all, child abuse is a widely recognized problem, and the number of people studying it must dwarf the number studying parental neglect in all other species combined. A student volunteered to check it out and came back disappointed. It was easy to find papers on child abuse, but they didn't seem to be asking the same questions or gathering the right kind of information to test even the most basic evolutionary hypotheses.

Martin and Margo might never have written their book but for two serendipitous events: Martin became allergic to mice and both were offered new jobs in their native Canada, at McMaster University in Hamilton, Ontario. On their way they drove through Detroit, Michigan, which had recently been dubbed "Murder City" by the press for having the highest homicide rate in the United States. It was then that they had their flash of insight about homicide statistics as a reliable source of information for testing evolutionary hypotheses. Martin needed a new study organism that didn't give him a rash—why not people? From this humble beginning, they gradually assembled information about murder from Detroit, modern and traditional societies around the world, and even back in time as early as thirteenth-century England to satisfy their "Show me the data" urge.

They soon discovered that even though nearly every death is investigated, the information gathered for each death often left much to be desired. Ingenuity would be required to make progress. Consider William Hamilton's theory of kin selection, which predicted that, all else being equal, animals should help (and avoid hurting) each other in direct proportion to their degree of genetic relatedness. This might seem obvious—a formal version of the adage "Blood is thicker than water"—but it was a major theoretical insight at the time, and Paul Sherman, whose work on pregnancy sickness and spices was described in the last chapter, initially made his name by showing that ground squirrels are more likely to warn a genetic relative about an approaching predator than a genetically unrelated neighbor. Fatal conflicts among kin are expected in some situations, as we have seen in the case of infanticide, but in general Hamilton's "Blood is thicker than water" rule should hold true. This was the first prediction that Martin and Margo hoped to test for humans with their homicide statistics.

To their surprise, they discovered that criminologists had come to a different conclusion. As two prominent experts on domestic violence in the United States put it, "With the exception of the police and the military, the family is perhaps the most violent social group, and the home the most violent social setting, in our society. A person is more likely to be hit or killed in his or her home by another family member than anywhere else or by anyone else."

This conclusion was based on the fact that roughly a quarter of the homicides in the United States take place among family members—but what counts as a family member? Crime statistics typically include a category such as "relative" for the relationship between the killer and victim, but this could be a spouse, an in-law, or a stepchild in addition to a genetic relative. Most crime statistics don't include these subcategories. When they do, spouses and marital relatives are far more likely to be the victims of homicide than blood relatives, as predicted by Hamilton's rule.

Another problem with the image of the family as a violent social setting requires a basic understanding of numbers versus ratios to appreciate. More people in New York City die in their beds every night than in Central Park. Does that make one's bedroom a more dangerous place than Central Park at night? Of course not, because many more people are *in* their beds than in Central Park. A ratio is needed to calculate risk—perhaps 1 death for 1,000 people in Central Park at night $(1/1,000 = .001)$ and 20 deaths for 8,000,000 people in their beds $(20/8,000,000 = .0000025)$. It might seem that professional criminologists would not make such a silly mistake, but Martin and Margo encountered it dismayingly often. Whenever they were able to calculate the appropriate. risk ratios, they discovered that blood was indeed thicker than water. Perhaps their most ingenious comparison involved collaborative homicides, where two or more people team up to kill a third person. If genetic relatedness makes no difference, then the average degree of relatedness between the killers and the victim should be the same as between the killers. Martin and Margo found appropriate data for testing this hypothesis in societies as diverse as tribal horticulturalists, medieval England, Mayan villages, and urban America. In each case, the killers were far more likely to be genetically related to each other than to their victim.

Daly and Wilson then turned to the specific subject of infanticide. Recall from Chapter 3 that infanticide has been studied in scores of animal species and tends to be associated with lack of resources, poor offspring quality, and uncertain parentage. Stepparents are certain that their stepchildren are genetically unrelated, which would seem to create a risky situation. Genetic relatedness is not the only relevant factor, of course, and many stepparents (including my own) have wonderful relationships with their stepchildren. Unlike Mexican free-tailed bat babies, human babies can be surrounded by love in addition to bodies and the circle of love can extend beyond blood relatives. Soon we will explore this special human ability, but for now we need to stress its partial nature. It would be naive to think that all stepparents regard their stepchildren in exactly the same way and with exactly the same concern as their own children. The average person without scientific training would scoff at such an idea and an evolutionist would be especially incredulous. Recall from Chapter 11 that the human mind is a mansion with many rooms, some so old that they precede our existence as a species or even the entire primate order. Female parental care is an ancient mammalian adaptation orchestrated by the same kind of unconscious "calculators" that I described for our eating habits in Chapter 8. Male parental care is much less common among mammals but still has evolved numerous times in response to environmental conditions that enable males to improve the fitness of their offspring by joining forces with the mother. Male parental care is also orchestrated by unconscious "calculators," despite its more sporadic occurrence. Even assuming that we are enlightened enough to consciously accord equal treatment to our stepchildren and natural children, it is hubris to think that our conscious decisions completely override these unconscious mechanisms. Statistically, it is a very good bet that children are more at risk from a stepparent than a natural parent. This was Martin and Margo's evolutionary prediction, but many would also regard it as just plain common sense.

Once again, Martin and Margo were surprised by what they encountered in the criminological and social scientific literature. Genetic relatedness was not recognized as a factor relevant to child abuse, and infanticide statistics often included only a category for "parent" without distinguishing stepparents from natural parents. When they were distinguished, there was the problem of numbers versus ratios. In one sample of fatal battered-baby cases from England, fifteen were killed by stepfathers and fourteen by natural fathers. These numbers provide the numerators but are meaningless until we provide the denominators—the number of babies that *reside* with natural fathers versus stepfathers, similar to the number of people in Central Park versus their bedrooms. Even though divorce and remarriage are common in England, as elsewhere, they seldom take place so rapidly that a small baby resides with a stepfather, at least compared to the number that reside with their natural father. Thus, the fourteen babies killed by their natural fathers must be divided by a very large number and the fifteen killed by their stepfathers by a much smaller number, precisely as with our Central Park example. When Martin and Margo calculated the ratios correctly, genetic relatedness emerged as not only *a* risk factor but by far the *most important* risk factor for infanticide. Depending upon the society from which they were able to find the necessary information, an infant is twenty to a hundred times more likely to be killed by a stepparent than a natural parent. In a graph where the height of a column indicates the risk of infanticide, the columns for stepparents and natural parents look like the Empire State Building standing next to a ranch house. It is important to remember, however, that even the highest rates are on the order of six hundred victims per million co-resident parent-child dyads per annum. Infanticide is not common and there is considerable variation among societies, even if stepparents evidently pose a greater risk than natural parents in all societies where information is available. There are many relevant comparisons to keep in mind.

These calculations don't even consider the possibility that the fathers categorized as "natural" might have had serious concerns about their actual paternity, the absence of resources, or offspring quality. In other words, the ranch house might have an evolutionary logic of its own, even if it is small standing next to the Empire State Building. With seemingly inexhaustible energy, Martin and Margo scoured the worldwide literature for traditional societies in addition to modern ones and showed that the "big three" factors explained the occurrence of infanticide in our species as well as any other species. There is also an important effect of maternal age. From an evolutionary perspective, parental investment decisions are fundamentally a trade-off between present and future reproduction. The older one gets, the less reproduction there is likely to be in the future. Infanticide should therefore occur most often in young women, which turns out to be true for societies as diverse as Canada and the Ayoreo Indians of South America.

This is only a fraction of what Martin and Margo accomplish in *Homicide*. I find their story richly ironic. They are studying not an arcane subject that only a scientist could love but one that is hugely relevant to human welfare. Politicians who like to ridicule scientific research might single out parental neglect in mice, but not parental neglect in humans. The government declares war on crime and funds an army of criminologists and social scientists to study it, as well it should. Even if this army shot at random, it seems that they should discover the major risk factors associated with a problem such as child abuse, whose most extreme manifestation is infanticide. Enter Martin and Margo, nonexperts armed only with predictions that are painfully obvious from an evolutionary perspective. Some are even obvious from a common-sense perspective, such as that children are more likely to be abused by a stepparent than a natural parent. Nevertheless, these predictions are so new that most of the experts aren't even prepared for them. Finding the relevant information is like sifting through mountains of gravel to find a few nuggets of gold.

"How has *Homicide* been received?" I asked them as we picked at our breakfasts.

"Some love it, but others hate it," Martin replied.

"Some become apoplectic," Margo added.

Given all the other ironies, this one no longer surprised me. Something is going on here that is more complicated than the typical rendering of the scientific process, something that enables the very same proposition to be self-evident to one person and inadmissible to another.

The morning HBES sessions were about to begin, and we needed to end our breakfast conversation. Martin and Margo wanted to catch a talk on mate choice, and I wanted to catch one on the use of horns as symbols in mythology. Martin is a bit of a hothead, however, and once he starts talking about the reception of *Homicide*, he finds it difficult to stop. The last words I heard him say as Margo pulled him through the lobby of the Hyatt Regency hotel were "Don't they know that *lives* are at stake?"

14

How I Learned to Stop Worrying and Love Genetic Determinism

SAY THE WORD "EVOLUTION" and some people hear the phrase "genetic determinism." If our behaviors are determined by our genes, and if our genes can't change, then it must be that our behaviors can't change, no matter how much we would like them to. Social injustices must be ineradicable because they are rooted in our genes. For many people, denying the human potential for change is the secular equivalent of denying the existence of God. The imagined implications of genetic determinism are so threatening that they lead to a form of secular creationism described in Chapter 1—the denial of evolution's relevance to human affairs, however essential for understanding the rest of life.

I count myself among those who cherish the thought that the world can be much better in the future, but I regard evolutionary theory as an essential tool for accomplishing change. Let me show you why the reasoning of the previous paragraph is flawed and why anyone who wishes to improve the human condition should want to become a sophisticated evolutionist.

No creature—not even a bacterium—is so simple that it is instructed by its genes to "do X." All creatures live in variable environments and are instructed by their genes to behave in a number of ways, depending upon the specific conditions that they encounter. At the very least, a caricature of genetic determinism would need to look like this:

In this situation ...	... behave this way
X	X'
Y	Y'
Z	Z'

We have encountered numerous examples of this kind of genetically determined flexibility in previous chapters. *If* you are a burying beetle with a small carcass, *then* reduce the size of your brood. *If* you are a female or a small male dung beetle, *then* don't develop a horn. *If* you are a young minnow and see a predator, *then* remain shy for the rest of your life. *If* you are an underweight fetus, *then* develop a thrifty metabolism. Behavior is strictly determined by genes in any particular environmental situation, as typically imagined for genetic determinism, but the existence of multiple situations adds a twist. Suppose that you regard behavior Z' as desirable. You could not achieve behavior Z' in situations X or Y no matter how hard you tried, but you could not *fail* to achieve behavior Z' in situation Z. Your plan of action for achieving desirable behavior Z' has become clear: supply environmental situation Z and you are done.

I don't mean to imply that improving the human condition is *that* easy, but there is something profound about this simple example that we need to savor before adding complications. As typically imagined, genetic determinism is threatening because it implies incapacity for change and the futility of environmental intervention. Envisioning genetic determinism as a set of if-then rules enables us to reach the exact opposite conclusion. The new genetic determinism provides a detailed recipe for change by specifying the right kind of environmental intervention. Anyone who hates the old genetic determinism should love the new one!

My example assumes that the species encountered environmental situation Z sufficiently often during its genetic evolution to evolve adaptive response Z'. Otherwise, there is no reason to expect Z' to be part of the behavioral repertoire, and we return to the dark implications typically associated with genetic determinism, as we have seen for the birds on the Galápagos Islands that mistook sailors for trees. On the other hand, my example doesn't include the fast-paced evolutionary processes built by genetic evolution that I described in Chapter 11 and will increasingly occupy our attention in future chapters. These can produce desirable behavior Z' in response to environmental situation Z, even if Z was not present during our genetic evolution.

An example will bring these abstract ideas to life. In 1997, the inexhaustible Margo Wilson and Martin Daly published an article in the *British Journal of Medicine* titled "Life Expectancy, Economic Inequality, Homicide, and Reproductive Timing in Chicago Neighborhoods." (I have reversed the order of their names because Martin is the first author of *Homicide* and Margo is first author of the article.) The city of Chicago is divided into seventy-seven neighborhoods for which homicide rates and other vital statistics are compiled separately. The neighborhoods vary tremendously in their quality of life, including the average length of life itself. Babies born into the best neighborhoods can expect to live *twenty years longer* (to their midseventies) than those born into the worst (their midfifties). Differences of this magnitude occur between developed and undeveloped nations, but Margo and Martin found them at a much smaller scale within a single American city.

In neighborhoods with the shortest life expectancy, women also tended to start having children at an earlier age. Teenage pregnancy is widely recognized as a social problem, but when ghetto women were asked why they were having children so early, they gave an answer that can only evoke sympathy. They said that they wanted their mothers to see their grandchildren and wanted to be around to see their own grandchildren. They used the word "weathering" to describe the deterioration of health that they observed all around them. If you and your loved ones were weathering at a fast rate, wouldn't *you* want to start having children early enough to see and help them raise their children?

Homicide rates varied tremendously among neighborhoods, from 1.3 to 156 per 100,000 per year. When Margo and Martin compared the homicide rate of a neighborhood with other causes of death (life expectancy at birth with homicide mortality removed), they found an amazingly strong correlation (r = −0.88). Most graphs of this sort show a fat cloud of points with an upward- or downward-sloping trend, but this one looks more like pearls on a string. Evidently, if you are a male born in the city of Chicago, the chance that you will commit or die from murder is very strongly related to your chance of dying from other causes.

It is perhaps more difficult to sympathize with inner-city males committing murder than with inner-city females having children, but consider the plight of a male who has almost no prospects for reproduction if he is a "nobody" and abundant prospects if he can become "somebody." Life has become a perverse lottery, with a few winners and many losers. Entering the lottery requires taking extreme risks to obtain status, including direct confrontations with other men who are competing for the same status. As the album and movie by rapper 50 Cent put it, "get rich or die tryin'." One of Martin and Margo's great accomplishments in *Homicide* was to explain the type of murder that seems to take place over nothing at all: two men get in an argument over a pool game or one insults the girlfriend of another and it escalates to a killing, often in full view of bystanders. Criminologists typically classify this type of homicide as "trivial altercation," which deeply misunderstands its basis. Martin and Margo show that it is about competition among males for status, which is anything but trivial.

Women seldom experience this degree of winner-take-all competition and almost never blow each other away in bars. Men only *sometimes* experience it, depending upon their social environment. It was easy for me to get married and have children. All I had to do was go to college and secure a good job. I am now fifty-six—past the age when the average inner-city Chicago man has died—and still in excellent health. I would never get into an escalating fight because I would have so much to lose and so little to gain. I am like the men in the best Chicago neighborhoods, who are no more likely than their wives to commit murder. I wish I could attribute my civilized behavior to my sterling character, but I have primarily my *situation* to thank. If I were to be suddenly transported into an inner-city neighborhood, I would almost certainly become one of the losers. If I had been transported at a much earlier age, I don't know how I would have fared, but I am certain that I would have tried to play a different game.

Status is inherently a relative concept. A status-conscious person will love his car if it is the best in town but hate the very same car if others are driving better ones. This fact enabled Margo and Martin to identify economic inequality as a factor that contributes to homicide rates, apart from average life expectancy or household income. Imagine plotting the household incomes of a single neighborhood on a graph, from poorest to richest. The line will be flat if everyone earns the same income and will become an ascending curve if people earn different incomes. The shape of the curve provides an index of economic inequality, sometimes called the Robin Hood index, regardless of the average degree of wealth. Margo and Martin calculated the Robin Hood index for each neighborhood and showed that it predicted homicide rates better than average income. People and especially men become dissatisfied when they perceive themselves as having less than others, regardless of how much they actually have. Economic inequality was already known to predict differences in homicide rates among nations (Canada versus the United States, for example). Margo and Martin showed that it could also explain differences at the much smaller scale of neighborhoods within a single city.

This wonderful study can be crudely represented by a list of if-then rules similar to the abstract list that began this chapter.

In this situation …	… behave this way
Unstable environment and low life expectancy	Take care of immediate needs and reproduce early
Stable environment and high life expectancy	Plan for the long term, including delayed reproduction
Winner-take-all status competition	Take extreme risks to obtain status
Reliable means for obtaining status without violent conflict	Avoid risk and work hard to obtain status

If we regard early pregnancy and coercive violence as problems, and if this list of if-then rules is roughly correct, then we have a clear plan of action: create a stable social environment with high life expectancy and reliable means for obtaining status without violent conflict, and we are done. If we fail to implement these environmental changes, we will probably fail no matter how hard we try, precisely as I described for my abstract example of genetic determinism at the beginning of this chapter.

You might be thinking that evolution isn't required to make such a reasonable prediction, but remember from the last chapter that a coherent theory was required to predict even something as reasonable as "Children are at greater risk from their stepparents than their natural parents." Here is how Margo and Martin describe the conventional scientific wisdom at the beginning of their article, which made their own approach worthy of publication in a top medical journal.

Psychologists, economists, and criminologists have found that young adults, poor people, and criminal offenders all tend to discount the future relatively steeply. Such tendencies have been called "impulsivity" and "short time horizons," or, more pejoratively, impatience, myopia, lack of self control, and incapacity to delay gratification. Behind the use of such terms lies a presumption that steep discounting is dysfunctional and that the appropriate weighting of present rewards against future investments is independent of life stage and socioeconomic circumstance.

There is an alternative view: adjustment of discount rates in relation to age and other variables is just what we should expect of an evolved psyche functioning normally. Steep discounting may be a "rational" response to information that indicates an uncertain or low probability of surviving to reap delayed benefits, for example, and "reckless" risk taking can be optimal when the expected profits from safer courses of action are negligible.

According to this assessment, conventional scientific wisdom has not converged on an image of the human organism as a set of if-then rules that would enable us to make positive social change spontaneous through the right kind of environmental intervention. That is why a basic understanding of evolution is needed to make such a reasonable image the new conventional wisdom.

The strongest form of genetic determinism would predict that each if-then rule is innately hardwired, as in my initial example. We might also expect innate sex differences. It is possible that males experienced winner-take-all competition so much more often than females during our genetic evolution that it comes far more naturally for them than for females. A weaker form of genetic determinism would predict that each if-then rule is determined by learning and cultural transmission, but that these are guided by psychological mechanisms that evolved by genetic evolution to arrive at biologically adaptive outcomes, at least most of the time. In this case, we could place women in a winner-take-all situation and expect them to start blowing each other away in bars, just like men. Both are reasonable possibilities, and scientific crank-turning will be required to determine the correct answer. Either way, there is no reason to be threatened by the prospect that men and women, on average, possess different if-then rules, as I will show in the next chapter.

A third possibility is that if-then rules can be created without any reference to biologically adaptive outcomes. In this anything-goes scenario, people are just as likely to plan for the long term as the short term in response to low life expectancy. Not only is this an unlikely possibility, but I don't see why anyone would *want* it to be likely. Secular creationists insist upon an anything-goes conception of human nature because they think it is required to establish their vision of the good life. Ironically, their own vision is better achieved by adopting a conception of human nature that fights tenaciously to survive and reproduce.

Evolution is inherently about organisms responding to environmental change. Only by the strangest of secondary assumptions can it be interpreted as denying the capacity for change. Religious and secular creationism have always been based on fear of the consequences of accepting evolution—why else would people who know so little about it feel so strongly about rejecting it? Once evolutionary theory is seen as a tool for positive change, it can be easily accepted, leading to insights that in retrospect appear like just plain common sense. As I said at the very beginning of this book, when it comes to evolution and its acceptance, the future can differ from the past.

15

They've Got Personality!

MY WIFE, ANNE B. Clark, is an evolutionist who works as hard as I do. We were graduate students at separate universities but met when we took the same tropical biology course in Costa Rica. There is nothing like falling in love in the tropics, especially when both of you are as entranced with the profusion of life all around you as with each other. At first it seemed like just one of those things, but then it seemed like just one of those things that should last forever. We married as soon as we obtained our Ph.D.'s, and two weeks later Anne left for South Africa, where she had a postdoctoral position to study a nocturnal primate called the thick-tailed bushbaby (*Otolemur crassicaudatus*). I had a postdoctoral position that was based at the University of Washington in Seattle, but it was mostly mind work that could be done anywhere, so I spent most of my time with Anne as she alternated between teaching at the University of the Witwatersrand in Johannesburg and her field site in the Northern Transvaal, near the border of what is now Zimbabwe and was then called Rhodesia. (It was there that I discovered that you can hear the sound of dung beetles coming before you can get your pants up.)

Anne's field site was a strip of forest that bordered a muddy stream running between the cattle pastures of a farm, whose owner kindly allowed her to work on his property. The largest African mammals were absent, but the diversity of wildlife was still astounding—aardvarks, bush pigs, civets, duikers, genets, guinea fowl, leopards, black and green mambas, mongooses, pythons, and vervet monkeys all inhabited the little riverine strip along with the bushbabies, while baboons barked from nearby cliffs. Almost nothing was known about bushbabies back then and high-tech methods such as radio tracking had not yet become common, so much of Anne's research involved tramping around at night making basic observations on individuals that had been trapped and marked by clipping and dying portions of their tails with Lady Clairol hair dye. We located the bushbabies with the help of red-filtered headlamps powered by motorcycle batteries that we wore on our hips. The eyes of nocturnal animals are highly reflective, and bushbaby eyes glow like fiery red coals even from a long distance. We became adept at identifying nocturnal animals by their eye shine. Two small circles closely spaced in the trees were a bushbaby. A single eye near the ground was a duiker (whose eyes are side-facing rather than front-facing). Two large widely spaced almond-shaped eyes near the ground was a leopard, something we seldom saw but I was always on the lookout for! A trailer parked next to the hut that we rented had two big reflector disks placed about four feet apart, which glowed brightly in the headlights of our car as we returned from a long night watching bushbabies. So obsessed was I with eye shine that my heart always skipped a beat at the thought of the enormous beast that those widely spaced eyes must represent!

The bushbabies soon became accustomed to the strange new beasts with the single red eye lumbering below them, just as they had become used to the lumbering cattle, but I must confess that I never became used to conducting fieldwork at night, worrying about leopards and black mambas in pursuit of bushbabies and their elusive eye shine. I often elected to stay in our hut, reading and scratching away at my equations, much to the disgust of the local farmers, who thought that I should be protecting my woman. Anne never seemed happier than in the company of her bushbabies, however, and would return at dawn with stories of their adventures. Her field study established that they have a richly complex social life, in contrast to their earlier reputation as primitive and solitary in comparison to the so-called higher primates.

One of the many things that Anne noticed about bushbabies was the striking differences in their personalities, even between members of the same family (they typically give birth to twins or triplets). In one family that she followed closely, one daughter was adventuresome, quickly made friends with other bushbabies, and ultimately established a home range that was separate from her mother's. Another daughter was almost always observed near her mother, appeared nervous in the company of other bushbabies, and ultimately remained in her mother's home range. Back in Johannesburg, Anne supported her field studies by offering novel objects to captive bushbabies. Some refused to come near, while others approached and explored the objects with confidence. In human terms we would call them shy and bold. Indeed, Anne's novel object experiments on bushbabies were patterned after experiments by the well-known psychologist Jerome Kagan, which revealed similar individual differences in human infants as young as six months old.

At that time (the 1970s) an entire branch of psychology was devoted to the study of human personality, but not from an evolutionary perspective. Biologists studying other species expected to find age and sex differences but usually didn't think to look for meaningful individual differences that cut across age and sex categories. In short, personality psychologists weren't thinking about evolution and evolutionary biologists weren't thinking about personality. Nevertheless, the idea of adaptive individual differences in behaviors such as boldness, movement, and sociability that cut across age and sex categories seemed eminently reasonable, not only for bushbabies but in general. While Anne was observing her behaviorally diverse bushbabies in the dead of night, I started to scratch down some equations by lamplight to show how natural selection might maintain a diversity of behavioral strategies within a single population.

After two years in South Africa, Anne and I moved to the University of California at Davis, where I had a job as assistant professor and Anne was provided an office as a courtesy. At that time, the academic world didn't know what to do with a husband and wife who were equally ambitious about their work. Jobs were scarce and offered one at a time. Getting two jobs at the same place was about as likely as winning the lottery twice. Old rules designed to discourage nepotism even prohibited a husband and wife from working in the same department at some universities. Our scientific colleagues at Davis welcomed Anne, but she was a second-class citizen as far as the university was concerned.

I started teaching and conducting research on burying beetles, as I described in Chapter 4. One of my field sites was a tract of land on the Pacific coast called Sea Ranch that had been developed into expensive properties. The Sea Ranch Association prided itself on ecological awareness and was pleased to allow a professor to conduct research on the grounds. As I wandered among the million-dollar condominiums with cans of rotten flesh to trap the beetles, I thought: "This is what the untouchable castes of India must feel like."

Anne made the best of her situation. She wrote papers based on her bushbaby research. Another thing that she had noticed about them, in addition to their diverse personalities, was that they gave birth to more sons than daughters. Her analysis was published in *Science* magazine and became a classic for the study of biased sex ratios in other species. She started new research on budgerigars, the familiar parakeets that you might well have as a pet, which are only a few generations removed from their wild Australian ancestors. They have stories to tell that might surprise you. Together we wrote a long article on why many species of birds begin to incubate their eggs before the last one is laid, causing their clutch to hatch asynchronously. You might think that this is an odd thing to focus upon, as arbitrary as a dog's curly tail, but it strongly reflects the shaping influence of natural selection. I was writing mathematical models of evolution in addition to my burying beetle work. Beetles on Monday, bushbabies on Tuesday, birds on Wednesday, equations on

Thursday. Her projects and mine, all mixed together from morning to night. Dozens of friends and colleagues at Davis pursuing their many projects, all part of one big glorious jigsaw puzzle being assembled with the help of evolutionary theory.

After three years we moved to the Kellogg Biological Station, a part of Michigan State University but located sixty miles away from the main campus. The station was created when W. K. Kellogg, the cereal king, donated his summer estate on the shore of Gull Lake to the university. I'll never be as rich as old man Kellogg, but I did get to enjoy his manicured grounds, half mile of waterfront, and Bavarian-style mansion as if I were a king. Away from the station, real estate was so cheap that we could buy an old, drafty farmhouse on seventy acres for the price of our California suburban ranch house with its postage-stamp yard.

Once again, I had the job and Anne had the courtesy appointment. I continued to work on burying beetles and mathematical models of evolution. Anne became interested in a flock of wild turkeys that roamed a bird sanctuary attached to the station. We both revived our interest in the idea of a single species as a collection of diverse individuals, as opposed to a uniform entity that inhabits a single "niche." Anne wrote a long review article with a graduate student named Tim Ehlinger, exploring the concept of personality as something that might apply to all species, not just humans. I published models showing that a single species could occupy multiple niches even when the individual specialists mated with each other and produced relatively inefficient generalists as offspring.

The lack of a real job was taking its toll on Anne, but life was good in other respects. We had our first daughter, Katie, followed by our second daughter, Tamar, five years later. We heated with wood, kept a big garden, and had chickens roaming the yard. We wandered the woods, fields, and marshes of our property. In fall we harvested apples from our derelict orchard to press for cider at a local mill. I still remember climbing the gnarled old trees, shaking their trunks, and laughing as the falling apples pelted my head and shoulders in the bright September sunshine.

I wondered whether having children would interfere with my work, but I didn't need to worry. When you love something you find a way to fit it in. Three a.m. might find me in the living room, pacing around with Katie to get her back to sleep, with her tiny body in one arm and a big book on the evolution of sex in the other. Both Katie and Tamar accompanied Anne to major conferences before they were two months old. As they grew older, they would troop after us like ducklings as we did our work. Our students were their best friends.

After a number of years I decided that I had studied burying beetles and their mites long enough. "I don't care if they are conceptually elegant," I thought. "They stink!" Besides, I had an idea for another project. Katie and I took some buckets and a couple of fishing poles and walked down the path through our woods to a lake that just touched our property. We paddled our canoe to the middle of the lake and spent a pleasant hour catching bluegill sunfish. Then we paddled into a shallow weed-choked bay and caught some more that we kept separate from the first batch. We took the fish to my laboratory, where Tim Ehlinger (the same student who wrote the paper with Anne) and I measured them carefully with calipers. We fed the measurements into a computer, crunched them with a statistical method called discriminate function analysis, and behold! The fish that we had captured in the open water and vegetation had different shapes. They were all members of the same species, living in the same lake, but the ones captured in the open water were shaped to cruise long distances and the ones captured in the vegetation were shaped to twist and turn through their spatially complex environment. I now had a fascinating example of a single species occupying more than one niche that I had been studying with my mathematical equations. Not bad for a day spent fishing!

After eight years at the Kellogg Biological Station and eleven long years without a proper job for Anne, Binghamton

University in New York State offered both of us gainful employment. Despite her hardships and neglect at the institutional level, Anne was so well respected among her colleagues that she was elected president of the Animal Behavior Society in 2003.

Not long after we arrived at Binghamton, a graduate student named Kristine Coleman enabled Anne and me to resume our mutual interest in animal personalities. Kris was a superb student who also happened to be extremely shy, speaking in a whispery voice when we could get her to speak at all. Would Kris like to conduct a pilot experiment on shyness and boldness in sunfish? Might they differ not only in their physical shapes but also in their behavioral responses to each other and their environments?

Our first experiment was simplicity itself, just like my fishing trip with Katie. Kris and I went to a pond containing pumpkinseed sunfish (a species closely related to the bluegill) and placed shiny metal minnow traps without bait at ten-meter intervals along the shore. It was fascinating to watch the behavior of the juvenile pumpkinseeds as these huge novel objects appeared with a violent splash and slowly descended to the bottom, like alien spacecraft descending to earth. At first the little fish scattered, but just as some people would flee from an alien spacecraft while others would gather to watch, at least some pumpkinseeds returned to explore the traps and entered the funnel-shaped openings. They were captured by their own curiosity. After ten minutes we removed the traps and used a seine (a long net) along the same stretch of shore to capture the fish that did *not* enter the traps. We were not comparing fish from different habitats in this case (open water and vegetation) but looking for behavioral differences between fish in a single habitat. If juvenile pumpkinseed sunfish vary along a shy-bold continuum, then the bold fish should differentially enter the traps. Sure enough, when we placed the trapped and seined fish in separate laboratory aquaria, the average trapped fish acclimated to its new environment and started feeding five days sooner than the average seined fish. Encouraged by this preliminary result, Kris spent the next four years studying shy and bold pumpkinseeds for her Ph.D. thesis.

In addition to conducting laboratory experiments, Kris needed a way to observe shy and bold fish in their natural environment, just as Anne tramped after her bushbabies in Africa. We figured out an ingenious way to do this using a square-shaped pond at Cornell University's experimental pond facility. We sank four corner posts and stretched a wire cable around the perimeter of the pond, which served as tracks for two cross wires attached to the perimeter cable by pulleys. We built a glass-bottomed observation chamber to float on the surface of the pond, just large enough for Kris to lie on her stomach and view the fish from above. The chamber was positioned at the intersection of the cross wires so that Kris could maneuver around the pond by pulling on the wires. Pulling on one wire would cause the other wire to move along with the chamber, thanks to the pulleys. The cross wires were marked at one-meter intervals so that the location of the chamber could be recorded at any time, like a dot on a giant sheet of graph paper. A lid was required to darken the chamber and prevent the fish from seeing Kris, making it look disturbingly like a floating coffin, but it provided a wonderful view of the fish. As far as they were concerned, the observation vessel was like a log floating aimlessly around the pond. Now we could trap and seine fish from the pond as in our pilot experiment, individually mark them with colored plastic beads attached to their backs like a body piercing, and return them to the pond for detailed observation. Kris spent many hours lying facedown in her floating coffin, getting to know pumpkinseed sunfish not as a homogenous population but as a community of individuals solving the challenges of feeding, avoiding predators, and interacting with each other in very different ways.

These first studies were voyages of discovery, but now the concept of personality has become a hot topic in animal behavior research. The conventional idea of a single species as a relatively uniform entity occupying a single niche has yielded to a much richer notion of a single species as a community of individuals employing different strategies to survive and reproduce. The strategies can involve mating, feeding, moving, socializing, or seeking protection from predators. They can become linked to age and sex or cut across these categories. Their proximate mechanisms can include genetic polymorphisms, developmental effects, or short-term behavioral flexibility. They are not limited to so-called higher animals but also exist in creatures such as insects, octopi, and even plants, which do not deserve the term "lower" once you get to know them well. In short, the diversity of life that is so amply reflected in the millions of species inhabiting the globe does not stop at the threshold of a single species but continues in the form of individual differences that we recognize with the vaguely defined term "personality."

Research has also become more ambitious and sophisticated. One species that is exceptionally well studied is the great tit—that's a common Eurasian bird, not the newest judging category of the Miss Universe pageant! Tits have been a favorite research organism for decades, providing an enormous amount of background information. They are easily studied in part because they nest in tree holes and readily accept nest boxes. In some study populations, hundreds of nest boxes are placed on the trees so that a large fraction of the tit population can be monitored. Adults can be easily captured in their nest boxes or at feeding stations. Their offspring can be measured and weighed on a daily basis. They can be easily and permanently marked with colored leg bands. Their behavior outside the nest boxes can be observed through binoculars during the day. These virtues have enabled a research group in the Netherlands, including Monica Verbeek, Piet Drent, and Niels Dingemanse, to learn more about the evolution of personality in the great tit than perhaps any other nonhuman species.

To describe just one of their experiments, parents and their nestlings were captured in their nest boxes and taken into the laboratory. The parents were released one by one into an observation room containing five artificial trees, each with four branches. They explored the room by flying between the trees and hopping among their branches. The amount of time required to visit four of the five trees was used as an index of exploratory behavior. A "fast" bird reached the fourth tree in less than one minute, and a "slow" bird could require more than ten minutes. The parents were returned to their capture site after testing, and the offspring were raised by the researchers to an age where they could be tested for their exploratory behavior. Offspring resembled their parents, suggesting but not proving that the trait is genetically heritable. Learning or shared environmental factors might also cause members of the same family to resemble each other. To distinguish among these possibilities, the fastest and slowest offspring were housed in separate aviaries with nest boxes so that they could reproduce, with due precautions to avoid inbreeding. Every time an egg was laid, it was replaced with a dummy egg and the real eggs were placed in the nest boxes of wild birds to be incubated. One day after hatching, the tiny nestlings were marked and mixed so that a single brood being raised by foster parents in the wild included offspring of both slow and fast birds in the aviaries. When the offspring were ten days old, they were retrieved from the nest boxes and returned to the laboratory, where the seemingly tireless researchers resumed their role as surrogate parents, raising the grand-offspring of the original adults to an age where they could be tested for their exploratory behavior. This procedure was continued for four generations. Just as artificial selection produced psychopathic and saintly chickens described in Chapter 5 and domesticated elite foxes in Chapter 6, it produced strains of superfast and superslow tits in this experiment, leaving no doubt that their exploratory behavior has a partially genetic basis.

This experiment required a heroic amount of work, but it was just a warm-up for the Dutch researchers. In subsequent experiments, they captured, marked, tested, and released over a thousand tits to study how fast and slow individuals interact with each other and their natural environment. They measured their survival, the number and condition of their offspring, how they forage and compete, how far they disperse from their natal nest, how long it takes them to recover from a startling event, and even how well they learn from each other. Fast individuals are perhaps best described as insensitive bullies who try to grab the best resources for themselves. Bullies who win have obvious advantages, but bullies who lose don't accept their fate very well. If tits held parties, some of the fasts would bore everyone with their loud conversation, while others would mutter into their beer that they don't get any respect. Slow individuals are best described as shy, sensitive types. They don't assert themselves, but they are observant and notice things that are invisible to the bullies. They are the writers and artists at the party who have interesting conversations out of earshot of the bullies. They are the inventors who figure out new ways to behave, while the bullies steal their patents by copying their behavior.

How these behavioral differences relate to survival and reproduction depends upon environmental conditions. Tits rely heavily upon beechnuts to survive the winter. Beech trees have evolved a clever strategy to protect their own offspring. They produce very few nuts for a number of years to reduce the population of nut predators and then flood the market with an abundant harvest. There are plenty of nesting territories to go around after a winter famine, but finding a territory after a winter feast can be as hard as finding an apartment in Manhattan. These flip-flopping environmental conditions alternately favor the fast and slow individuals, like a suspenseful basketball game in which the lead keeps changing. A winter famine favors fast females but a housing glut favors slow males. Fast males get the best territories but slow females are more attentive mothers. If you are a tit, your fitness can even depend upon the personality of your mate. In one year, fast-fast and slow-slow pairs were more successful than mixed pairs, who evidently had compatibility issues.

Even after this heroic amount of work, the Dutch researchers would be the first to admit that tit personalities are more complex than a one-dimensional continuum from fast to slow, but this only reinforces the main message of their research: there is no single best tit personality but rather a diversity of personalities maintained by natural selection. Moreover, variation at the behavioral level is based in part on underlying genetic variation. In principle it might be possible for behavioral variation to be caused entirely by mechanisms such as developmental switches or short-term learning, without any underlying genetic variation, but that is not the case for the great tit.

These wonderful studies of animal personalities, the vast majority conducted within the last twenty years, provide a panoramic background for thinking about human behavioral diversity as a community of strategies that solve the problems of survival and reproduction in different ways. A recent study of human extroverts and introverts by psychologist Daniel Nettle is strikingly similar to the research on fast and slow tits. Extroverts actively seek stimulation and novelty while introverts tend to be more reserved, especially in unfamiliar situations. If we were to release extroverts and introverts into a big observation room and measure their exploratory behavior, they would almost certainly be identified as fast and slow. Nettle did not perform this experiment but merely administered a standard personality test measuring extroversion to a large number of adults and asked them some additional questions about their lives that are obvious from an evolutionary perspective but had not occurred to psychologists unfamiliar with natural selection thinking. Extroverts had a constellation of grab-life-by-the-horns attitudes. They had a greater interest in sex, were more ambitious and competitive, and enjoyed travel more than introverts. It is easy to imagine some of them holding court at a party and others muttering into their beer that they don't get any respect. These attitudes were reflected in their self-reported life histories. There was a strong relationship between extroversion and lifetime number of sexual partners for both men and women. Extroverts tended to end their own romantic relationships to begin another, whereas introverts tended to have their romantic relationships ended for them. On the other hand, extroverts were also more likely to be hospitalized for an accident or illness. When you grab life by the horns, sometimes you get gored.

The benefits and costs of introversion are explored in a

comprehensive review article by psychologists Elaine and Arthur Aron. They argue that a fundamental axis of variation in both humans and other species involves the processing of information. Information is a mixed blessing; too little can be disastrous, but too much can be overwhelming. There is no single best solution to this trade-off. Sometimes it can be best to follow an inflexible routine, and at other times it is better to consider all of one's options. Individuals have some ability to toggle between these alternative strategies, but there are also individual differences, which are almost certainly either genetically based or triggered very early in development. A nervous system designed to process lots of information simply must be different from one that forges ahead inattentively. Highly sensitive people (HSP), as the Arons call them, can't avoid processing information. Indeed, their sensitivity appears to be quite general, including pain, bright lights, coarse fabrics, loud noises, and drugs in addition to mental processing. A person who reports having a rich, complex inner life and being deeply moved by the arts also tends to report being sensitive to caffeine and startling easily.

Highly sensitive people (or animals) are likely to be slow in a novel situation for the simple reason that they are processing new information. They are "stopping to check it out" rather than "forging ahead." Given too much sensory input, highly sensitive people tend to become overwhelmed and withdraw from the situation. This is a form of shyness, but highly sensitive people are not intrinsically shy. After all, the purpose of processing all of that information is to arrive at new solutions to life's problems. A highly sensitive person who succeeds at doing this can become as outgoing and gregarious as someone who is merely "forging ahead." Some individual differences (such as sociability) are not innate but are manifestations of other individual differences (such as information processing) that are.

It might seem that highly sensitive people are less resilient than those who forge ahead, but the very opposite can be true under certain circumstances. The Arons begin their article with the following passage by Victor Frankl, the world-renowned psychiatrist and Holocaust survivor:

Sensitive people ... may have suffered much pain [in the concentration camps] (they were often of a delicate constitution), but the damage to their inner selves was less. They were able to retreat ... to a life of inner riches and spiritual freedom. Only in this way can one explain the apparent paradox that some prisoners of a less hardy make-up often seemed to survive camp life better than did those of a more robust nature.

So much has happened since Anne was struck by the distinct personalities of her bushbabies, reflecting the light of her headlamp with their fiery red eyes. Not only are single species recognized as diverse communities in their own right, but our own species has been added to the list. Psychologists such as Daniel Nettle and Elaine and Arthur Aron cite the animal literature and employ the same evolutionary reasoning as Kris Coleman staring down at her fish and the Dutch researchers staring up at their birds. Darwin would be pleased, although perhaps he would wonder why it took us so long to seamlessly combine the study of ourselves along with the rest of life, as he was able to do so long ago.

Should we feel threatened by the prospect that some aspects of our personality are defined by our genes or very early in our development—that one person might find it difficult to take life by the horns and another might find it equally difficult to be overwhelmed by a symphony? I think I know how Elaine Aron would answer this question. She is a practicing psychotherapist in addition to her academic career and has gathered a large following of highly sensitive people through her books, workshops, and Web site (http://www.hsperson. com), where you can take a simple test to discover if you are a highly sensitive person. In plain language that anyone can understand, she explains that the trait is normal, is present in about 15 to 20 percent of the population, and exists in about the same proportion in other species. It is a great gift that can also be a liability in some situations. It is especially misunderstood in our own culture, which values toughness and regards extreme sensitivity as abnormal (it's easier to be a highly sensitive person in Asia). Readers who are quoted at the beginning of her book *The Highly Sensitive Person* were anything but threatened: "You put into clear and understandable words what I have always known about myself." "I cannot thank you enough for the inner peace your book has given me!" "It has really made me feel like part of a larger group, and not quite so weird after all."

I find it even more affirming to contemplate that this "larger group" includes not only highly sensitive people but highly sensitive creatures, great and small.

The Beauty of Abraham Lincoln

A SENSE OF BEAUTY has an ethereal quality that seems to occupy a different plane of existence from the raw struggle to survive and reproduce. Darwin's theory might explain why we grub for food and fight for mates but not why we gaze appreciatively at the sunset. It might explain why we make bowls but not why we decorate them.

In fact, an evolutionary theory of aesthetics is emerging that promises to explain not only our own sense of beauty but also a kindred sense in many other creatures. We have already seen that even so-called simple species such as insects have sophisticated "calculators" for solving their particular problems of survival and reproduction, however incapable they are of solving novel problems. What would such a calculator look like for, say, a coyote on a hilltop surveying the vista below, trying to decide which way to go? As it gazes from left to right, mental processes would need to assess and compare all of the features relevant to survival and reproduction—the presence or likelihood of food and water, places that offer protection from predators and the elements, and so on. These calculations need not be conscious. They are more likely to take place automatically beneath conscious awareness, like breathing, seeing, and hearing. Just as the coyote doesn't know that its mind is converting the light striking the two-dimensional surface of its retina into an elaborate three-dimensional representation of the world, it also doesn't know that its mind is computing the best direction to head down the hill. All it experiences is a magnetic attraction toward the unoccupied cave amidst lush vegetation with the rabbits hopping about, which seem beautiful.

In short, an evolutionary theory of aesthetics is based on three claims: (1) all creatures have evolved to assess their environments to make adaptive choices, (2) the mechanisms of assessment often take place beneath conscious awareness, and (3) the mechanisms are subjectively experienced as a feeling of attraction toward features of the environment that enhance fitness (beauty) and repulsion from features that reduce fitness (ugliness). If this theory is even partially correct, then our sense of beauty, like our personalities, can be studied as something continuous with the rest of life rather than forever remaining an enigma from an evolutionary perspective.

Right or wrong, at least some aspects of the theory are testable. Is there or isn't there a correspondence between what we regard as beautiful and what enhances our biological fitness? As soon as we begin asking this question, either scientifically or on the basis of common experience, a remarkable transformation takes place. Some observations remain enigmatic, such as gazing at a sunset or decorating a bowl, but others fall into place with such a resounding thud that they appear obvious in retrospect. Consider our taste for landscapes. Most of us love to surround ourselves with water, lush vegetation, and open spaces dotted with trees. We even find landscapes more beautiful when they include large animals grazing! We enjoy living in houses that make us feel protected while providing a view outward. We delight in the warmth of a crackling fire. These feelings run so deep that hospital patients recover faster in a room with a window and even faster if the window includes a pleasant view. You might wonder why it is a custom to bring flowers to sick people, but it turns out that having flowers in one's room also speeds recovery from illness. All of these features of our surroundings that give us joy and strike us as beautiful are clearly related to biological survival and reproduction. Take them away and we become stressed, even to the point of impairing our health.

The theory also makes sense of why individuals and cultures often differ in their perception of beauty. Suppose that you are a teenager from a tiny rural town that holds nothing for your future, at least as far as you can see. It can seem hideously ugly compared to the bright lights of the city. In Paul Simon's song "My Little Town," even the colors of the rainbow turn black. The same town can appear heartrendingly beautiful to city dwellers who have made their fortune and want to get away from it all. Most of the American pioneers regarded the wilderness as ugly and couldn't wait to cut it down to make way for their beautiful homesteads. The same wilderness—or what little is left of it—is worshipped today for being "unspoiled." All of these disparate perceptions of beauty have one thing in common—*that which is regarded as valuable is also regarded as beautiful.*

As for the beauty of our surroundings, so also for our perception of human beauty. A clear complexion, strong teeth, and lustrous hair are excellent indicators of both health and beauty. From women ogling the men on muscle beach to men drooling over *Playboy* centerfolds, what is it that we don't understand about beauty as an unconscious assessment of the fitness value of our potential mates? If we wish to go beyond common experience, evolutionary psychologists love this topic and have been spinning the wheel of science with gusto. To pick a recent example, a team of British researchers headed by S. Craig Roberts took a blood sample from ninety-seven men and also photographed their faces under standard lighting conditions. The blood was used to measure heterozygosity (the presence of different alleles, or versions of a gene, at a given site on the chromosome) for three genes in the major histocompatability complex (MHC), which enables our immune system to distinguish between our own cells and the cells of invading disease organisms. In general, the more our MHC genes are heterozygotic, the healthier we are likely to be. The photographs were digitally cropped so that only the face was visible and then were presented in random order to fifty women, who rated their attractiveness on a seven-point scale. The faces were rated as physically attractive in direct proportion to the number of genes that were heterozygotic. The researchers then digitally excised a patch of skin from each image between the nose and the upper lip, enlarged it by 300 percent, and asked the women to evaluate the apparent healthiness of the skin patch. Again, their ratings corresponded to the number of MHC genes that were heterozygotic. They didn't even need to see the rest of the face. It is as if the women were giving the men a detailed medical examination with their eyes and reporting a clean bill of health with their assessment of attractiveness!

Faced with the obviousness-in-retrospect that beauty is partly an unconscious assessment of fitness value, it is easy to conclude that nothing new has been accomplished. *Of course* we like white teeth better than rotten teeth and clear skin better than festering sores. *Of course* we like water and lush vegetation better than a parched desert. Anyone in a bar or a supermarket could tell us that, so why should we waste tax dollars to study it scientifically? Moreover, there is something vulgar about these forms of beauty, as if they account for the appeal of calendar art and pornography but not the ineffable qualities of "great" art or the "true beauty" of a person, whatever these might be.

Even if our evolutionary theory of aesthetics never gets beyond calendar art and pornography, it is important to assert that something new *has* been accomplished. My dictionary defines vulgar as "ordinary, commonplace; not advanced or sophisticated; offensively coarse in manner or character; low." It is one thing to claim that aesthetics is uniquely human and has nothing to do with survival and reproduction. It is quite another thing to claim that aspects of aesthetics regarded as "commonplace" are easy to explain in terms of survival and reproduction, leaving something left over that we call "sophisticated." As for being obvious in retrospect, nothing is less obvious than the concept of obvious. Whenever Sherlock Holmes explained the reasoning behind his amazing deductions, Watson exclaimed, "That's obvious!" much to the annoyance of his brainy friend. The same observation can be obviously true, obviously false, or just plain invisible, depending upon the background of other observations and assumptions. What is obvious today was not obvious a hundred years ago, and what was obvious then is hard for us to comprehend. I began this book by saying that learning about evolution can be like walking through a door and not wanting to return. It can become as natural as riding a bicycle, which of course is not natural at all but merely second nature. For those who experience this transition, much becomes obvious that was not obvious before.

Moreover, our evolutionary theory of aesthetics can take us beyond calendar art and pornography. Ask people—even "common" people—what they value most in a long-term mate and the top qualities are usually nonphysical, such as niceness, intelligence, and, intriguingly, a sense of humor. Physical qualities such as good looks and material qualities such as wealth tend to be ranked lower, with large individual differences and more modest sex differences. In one study, for example, good looks received an average ranking of 3 for men and 6 for women, while wealth received an average ranking of 8 for women and 11 for men. The importance of nonphysical qualities in a social partner, whether a mate, friend, or associate, makes perfect sense from a purely biological perspective.

It follows that the perception of beauty should be influenced by nonphysical factors in addition to physical ones. Consider Sam, who ponders whether to marry Jenny. Her physical traits are only average, but she is exceptionally nice and intelligent and has a marvelous sense of humor. Perhaps he keeps her physical and nonphysical traits separate in his mind. He might think (consciously or unconsciously), "Jen only rates a 5 for looks but is so wonderful in other respects. I think I'll propose!" On the other hand, perhaps his mind merges her physical and nonphysical traits into an overall estimate of fitness value, which is perceived as beauty. He is simply drawn to her, including her looks, and would rate her as more physically attractive than others who are unaware of her nonphysical virtues. Both of these scenarios are possible—remember that a theory can do no better than suggest plausible alternatives—but the second is most faithful to the concept of beauty as an unconscious assessment of fitness value.

I became fascinated by this possibility a number of years ago and decided to test it with the help of one of my graduate students named Kevin Kniffin. Kevin was the first member of his family to attend college, and a full merit scholarship at Lehigh University enabled him to follow his muse without worrying about repaying a massive debt. He gravitated toward the big questions, such as why people cooperate, and came under the influence of two professors, John Gatewood and Donald Campbell, who addressed these questions from an evolutionary perspective. At first he intended to become a gypsy after graduating (as he describes it), rambling around the country with no particular goal in mind, but a basketball injury forced him to remain at home for a year, where he read more about evolution from books gleaned from Philadelphia's used-book stores. By then he was sufficiently intrigued to attend graduate school to work with me on human groups. His interest in sports led him to consider sports teams as modern-day cooperative units, and for his master's thesis he was studying the university's crew team the way anthropologists study distant tribes—by living among them and observing their every move. No deception was involved—the coach and team members knew what he was doing—but they also accepted him as a teammate who shared the pain of their grueling workouts, attended the meets, and so on. The beauty project was new for both of us. We knew that we were too busy, but it was too interesting to pass up. Projects are like children—so much fun to conceive that you forget how much work is required to bring them to maturation!

We began in a lighthearted fashion with a study involving the rating of photographs. Hundreds of studies of physical attractiveness have involved rating the photographs of strangers, similar to the study of MHC genes described earlier. We added a twist by having people rate the photographs of people they knew—their classmates in their high school yearbooks—for familiarity, liking, respect, and physical attractiveness. Then we had the photographs rated for physical attractiveness by another person of the same age and sex as the yearbook owner, who did not know the people in the photographs. If the perception of physical attractiveness is based on purely physical traits (visible from the photographs), then the two raters should largely agree with each other, regardless of what one knows about their nonphysical qualities. If the perception of physical attractiveness is based on an assessment of overall fitness value, then the yearbook owners' rating of physical attractiveness should be influenced as much by what they know about the people in the photographs as by the second raters' assessment of purely physical traits.

The first person to participate in this study was our department secretary, a kindly woman in her fifties who still had her high school yearbook. Her assessment of physical attractiveness corresponded much more strongly to how well she liked her classmates than the assessment of the second rater. Fascinated, I showed her the photograph of the person that she rated as least physically attractive. It was a male who looked completely average to me and to the second rater. At the mere sight of his photograph, her face became wreathed in disgust as she told me about what a horrible person he was, what a foul mouth he had, and other failings of character that had nothing to do with his physical features. His character had become imprinted upon his physical features so strongly that she acted as if she were ejecting rotten food from her mouth, even though she had not seen or interacted with him for over thirty years!

Encouraged, we enlarged our sample size to twenty-seven yearbook owners who also functioned as second raters by evaluating the photographs of someone else's yearbook in addition to their own. Nonphysical qualities had a massive influence on the assessment of physical attractiveness. As predicted by the theory, it is not how well you *know* a person (familiarity per se) but how well you *regard* him or her (liking and respect) that makes the biggest difference. There were also sex differences and individual differences within each sex. Men tended to base their assessment of physical attractiveness on purely physical traits (agreeing with the second rater) more than women, but this was just an average difference; some men appeared to be influenced almost entirely by liking and some women almost entirely by physical traits. It would be fascinating to relate individual differences in the assessment of beauty to individual differences in personality discussed in the previous chapter, but that is a study for the future. Incidentally, this study is so easy to do that it would make a good class exercise. Simply have the students bring their yearbooks to class and the data can be collected during a single period. My daughter Tamar and her friend Nicole Benson did a version of the experiment as a project for their high school statistics class and confirmed our results—plus garnered an A for the project.

Fortuitously, an unfolding drama among members of the crew team that Kevin was studying for his thesis became relevant to our study of beauty. A member of the men's squad had emerged as a slacker, often showing up late for the grueling practices and sometimes even failing to show up at all. Rowing requires mutual effort, and this was a serious violation of the social contract. The slacker became the target of malicious gossip (something else that Kevin was studying), and the hardest-working members of the team were praised by comparison. Might this difference in social standing be reflected in the assessment of physical attractiveness? Kevin asked the team members (both men and women) to rate each other for talent, effort, liking, respect, and physical attractiveness. Then he had photographs rated by strangers, as in our previous study. Sure enough, the nonphysical characteristics were highly correlated with each other and with the assessment of physical attractiveness among the team members and not at all with the strangers' assessment of physical attractiveness from photographs. The slacker had become ugly, but only to those who knew him.

These two studies were encouraging but also had some weaknesses. Perhaps a single photograph does not provide enough information about purely physical traits compared to seeing the person in the flesh. Perhaps two separate raters disagree about what counts as physically attractive even on the basis of purely physical traits. The ideal study would have a group of people rate each other for physical attractiveness after meeting in the flesh for the first time and then again after getting to know each other's nonphysical qualities. Fortunately, a summer archeology class provided just this opportunity. Students worked on a dig site eight hours a day, five days a week, for a six-week period. With the permission of the instructor and the university's review board (required for all research on humans), Kevin visited the class on the first day and had the students rate each other on a scale from 1 (low) to 9 (high) for familiarity, intelligence, effort, liking, and physical attractiveness. Since some of these questions were inappropriate on the first day of class, the students were told to answer even if they had only a vague impression or to leave their answer blank if they had no impression at all. Name panels were placed by each student's seat to facilitate identification. The same questionnaire was completed during the last day of class. Kevin also spent several days working with the class and interviewed the instructors to obtain a descriptive account of the group and the participation of its members. Sure enough, the students changed *their own initial assessment* of physical attractiveness after getting to know each other. One woman who received an average score of 3.25 for physical attractiveness at the beginning proved to be a hardworking and popular member of the class. Not only was she well liked, but her physical attractiveness rating soared to 7.00!

My favorite example of a person made beautiful by his nonphysical qualities is Abraham Lincoln. In his lifetime he was regarded as hideously ugly, especially by his political opponents, who compared him to a gorilla. Even Lincoln made fun of his appearance. Once, when accused of being two-faced, he replied, "If I had two faces, do you think I would be wearing this one?" Yet it is impossible for most of us today to look upon his face without an upwelling of love and devotion. We don't say, "What a wonderful man—too bad he looked like a gorilla." We love his face, which has become inseparable from his admirable qualities. Our evolutionary theory of aesthetics can potentially explain not only the "vulgar" beauty of calendar art and pornography but even the ineffable beauty of Abe Lincoln.

I will take up the question of why we make objects of beauty (art) in Chapter 24 and the intriguing question of why we value a sense of humor in Chapter 23. For now, at the very least I can offer an unusual beauty tip: if you want to become more *physically* attractive, become a more valuable social partner. Ignore this advice and no matter how hard you work on your outward appearance, you might become like the victim of a remark I overheard one woman make to another on their way across campus: "If I didn't know him and hate him, I'd think he was cute!"

Love Thy Neighbor Microbe

IN THE LAST TWO chapters I attempted to show that what seems distinctively human, such as our diverse personalities and sense of beauty, is deeply continuous with the rest of life. Now I will attempt to make the same point for morality and religion. This will require a number of chapters, as befits the magnitude of the subject. Religions frequently portray the conflict between good and evil as a struggle of cosmic proportions. It turns out that they are right. The conflict is eternal and encompasses virtually all species on earth. Although it's difficult to check, I would bet everything I own that the conflict exists on any other planet in the cosmos that supports life, because it is predicted at such a fundamental level by evolutionary theory.

We already began our journey in Chapter 5, where I showed that morally laden terms such as "good" and "evil" have a surprisingly simple biological interpretation. Traits associated with "good" cause groups to function well as units, while traits associated with "evil" favor individuals at the expense of their groups. I also hinted that groups whose members are as good as gold toward each other can behave toward other groups in the same way that evil individuals behave toward members of their own group. There appear to be levels of good and evil, an observation that will occur again and again in the chapters that follow.

I made these points not by forcing my own interpretation of good and evil upon you but by asking my students what *they* associate with good and evil. It was *their* lists, not mine, that were so easy to interpret in terms of the functioning and undermining of groups. I would love to repeat the exercise in different cultures around the world. In fact, if you belong to a culture different from mine, please take a survey among your friends and send me the results. I suspect that individuals and cultures will disagree about who belongs inside the moral circle (family, tribe, nation, men only, humanity, all species) and what it takes for a moral group to function well (such as conformity versus tolerance of differences) but that nearly everyone will associate morality with the good of the group defined by the moral circle and immorality with the undermining of the group.

In the absence of a cross-cultural survey, we can consult religious traditions around the world and throughout history. According to St. Gregory the Great in the late sixth century, the seven deadly sins are (from least to most serious) lust, gluttony, avarice, sloth, anger, envy, and pride. Some of these might sound strange and even amusing to members of modern affluent societies who worship at the altar of consumerism, but they make clear sense against the background of the times. Dante's definition of pride, for example, was "love of self perverted to hatred and contempt for one's neighbor." When Rabbi Hillel was challenged to explain the Torah in the time that he could stand on one foot, he famously replied, "That which is hateful to you, do not do to your neighbor ... the rest is commentary." All of the great religious traditions are said to embody the Golden Rule (in plain language, "Treat others as you want to be treated"). The British anthropologist E. E. Evans-Pritchard called "the elimination of the self, the denial of individuality, its having no meaning, or even existence, save as part of something greater, and other than the self" a "psychological fundamental" of so-called primitive religions.

Philosophers might cringe at my reliance upon lay and

religious conceptions of morality, as if they have nothing to say about the subject. What about the problems associated with utilitarianism, Kant's categorical imperative, and all of those carefully crafted moral dilemmas that dumbfound our intuition? As with our sense of beauty, there seems to be a common sense of morality captured by the Golden Rule and something else that we call sophisticated. As with our analysis of beauty, we need to begin by explaining the moral equivalent of calendar art and noting our accomplishment if we succeed. It is one thing to claim that morality is uniquely human and has nothing to do with evolution. It is quite another thing to claim that aspects of morality regarded as "common" are easy to explain in terms of evolution, leaving something left over that we call "sophisticated."

It is the "do unto others" form of good and evil that can be found whenever there is life. The simplest form of life on earth is a virus, which can afford to be simple because it parasitizes the much more complex machinery of free-living organisms. When a virus enters a cell, its genes instruct the genes of the cell to make products that can be assembled into more viruses. These products diffuse throughout the cell, creating a kind of broth for viral replication. Eventually the cell bursts, releasing hundreds of viral progeny to search for and invade other cells.

Occasionally it happens that a viral particle becomes an ineffective parasite by losing some of its genes. It can no longer cause the cell to *make* the products for its replication, but it can still *use* the products provided by other viral particles in the same cell. In fact, its shortened genome can replicate faster than genomes of normal length. In economic terms, the normal virus is providing a public good by contributing to the broth of gene products that can be used by all viruses within the cell. The mutant virus is "cheating" by profiting from the public good without contributing to it. It is guilty of the fourth deadly sin, sloth.

A cheating virus particle benefits at the expense of its solid-citizen neighbors within the same cell, but what happens when its progeny are released into the environment to seek new cells? They can continue to profit from their deadbeat ways if they enter cells that also contain solid citizens, but they are out of luck if they enter cells by themselves. Their long-term success therefore depends upon the average number of viral particles that infect a given cell. Biologists call this "co-infection," and it can be easily manipulated in the laboratory. Decrease the concentration of viral particles, and the majority of cells will be infected by a single particle. The cheaters will have no one to exploit and will decline to a very low frequency in the population. Increase the concentration of viral particles, and the majority of cells will be infected by many particles. The cheaters will usually have solid citizens to exploit and will increase to a high frequency in the population. Like the Greek gods on Mount Olympus, who were frequently indifferent to and even spiteful toward the mortals below, we can alter the tide of battle between good and evil, sloth and industry, merely by altering the concentration of viral particles in the test tube.

Bacteria are among the simplest free-living organisms. In Chapter 8, I described how a bowl of soup left on your windowsill is like a remote island such as Hawaii as far as bacterial evolution is concerned. With their ultrashort generation times, the first bacteria to colonize the "island" diversify within a matter of days to occupy a number of ecological niches. Paul Rainey, a microbiologist who divides his time between Oxford University in England and the University of Auckland in New Zealand, studies bacterial evolution under carefully controlled laboratory conditions. He provides the "soup" in the form of a sterile liquid culture medium and introduces a single bacterial species, *Pseudomonas fluorescens*. The population grows rapidly until it becomes starved for oxygen, creating an advantage for a mutant type that can form a mat on the surface, providing access to oxygen from above and nutrients from below. The ancestral form is called "smooth" (SM) and the mutant form is called "wrinkly spreader" (WS), based on the appearance of their colonies when grown on agar plates.

WS creates a mat by producing copious amounts of an adhesive polymer that sticks the bacteria to each other. The polymer is expensive to make, setting the stage for another social dilemma. Mutations within the mat create deadbeats whose motto is "Make babies, not glue!" The deadbeats spread at the expense of the solid citizens until the mat disintegrates and everyone falls to the bottom of the bowl, like the lost city of Atlantis. Thus is the glue of civilization dissolved by sloth!

These two examples of good and evil among microbes can be repeated without end because they are based on inescapable facts of social life. Individuals often live in groups, whether they want to or not, and their efforts on their own behalf are vulnerable to exploitation, as in our viral example. Individuals can often work together to create something greater than what anyone could have achieved on their own, as in our bacterial example, but the joint effort is again vulnerable to exploitation by individuals who share the benefits without sharing the cost. These problems will exist wherever there is social life. They are perceived as moral problems in our own species but as evolutionists we need to understand industry and sloth, good and evil, as alternative strategies that succeed and fail in different ways. This is not an abandonment of morality but a prelude to thinking about the nature of morality from an evolutionary perspective. For the moment, we need to adopt the detached perspective of the Greek gods, watching the comedy and tragedy of life below. There are the deadbeats, taking advantage of the solid citizens within each group. There are solid citizens that function well as groups, at least until they are corrupted by the deadly sins. The struggle is eternal, although we can tilt the advantage according to our whim.

Perhaps the most impressive example of microbial virtue is the cellular slime mold *Dictyostelium discoideum*, affectionately nicknamed Dicty by the hundreds of scientists who study it. Dicty is among an elite group of "model organisms" that have been selected by the scientific community for focused research. The idea is to study a small number of species in extreme detail to elucidate basic life processes common to many species. As a model organism, Dicty has received vastly more attention than it would on the basis of its fascinating natural history. The full arsenal of modern scientific technology has been brought to bear upon it, from gene sequencing to clever methods of visualizing chemical interactions taking place inside a single cell. You can marvel at some of these techniques for yourself by visiting dictyBase, a Web site where Dicty scientists get together to swap information, pictures, and more (http://dictybase.org). This might seem like the ultimate in nerdiness, but try it once and I'll bet that you come back for more!

Dicty is an amoeba, a single-celled creature that continuously changes its shape, moving and engulfing its food by sending out blobby extensions of itself called pseudopods (false feet). You can see videos of single amoebae on the Web site, including a remarkable side view taken with a confocal microscope, which uses laser beams and a lot of computer processing to construct three-dimensional images of tiny objects. Dicty is not just any amoeba. When other species of amoeba run out of food or moisture, they form into protective capsules called cysts until conditions improve. Dicty responds by calling upon its neighbors for help. Specifically, it releases cyclic AMP (cAMP), a molecule that is also important in many intracellular processes. In addition to releasing cAMP, the surface of each Dicty is studded with thousands of receptors that are sensitive to cAMP. When a Dicty encounters a concentration gradient of cAMP, its receptors are stimulated more on one end of its tiny body than the other, indicating the direction to head toward its neighbor.

Since a Dicty in trouble is at the center of its own cAMP concentration gradient, you might wonder how it can detect the gradient produced by a neighbor. Ingeniously, it produces its own cAMP in a pulsed fashion, which diffuses away from its body in a spiral wave. Its receptors are synchronized to become sensitive to cAMP only during the troughs of its own spiral wave, enabling it to sense the waves emanating from neighbors. Dicty scientists have developed ingenious ways of visualizing both the waves and the receptor sites in video clips that you can view on the Web site. The waves spiral out from each individual like computer art and the receptors light up like miniature headlights, steering each Dicty toward its neighbors. As they begin to aggregate in clumps, they synchronize their signals to produce a larger spiral wave that looks for all the world like the rotating beam of a lighthouse. Smaller clusters are attracted to larger clusters, culminating in enormous aggregations of up to a hundred thousand individuals.

Now that the clan has gathered, it transforms itself into a sluglike object consisting of the cells embedded in a gelatinous matrix that they have secreted. The slug is capable of moving distances up to twenty centimeters. Considering the size of a single Dicty, this is like you traveling forty miles. Not only can the slug move, but it can actually see well enough to move toward light! How these marvels of coordination are accomplished has been worked out in equally marvelous detail by Dicty scientists. In one video clip, the cells in the slugs have been tagged with fluorescent dyes that glow brightly in different colors, depending upon the stage of cell development. The front end of the slug is green, the back end is orange, and all of the cells are rotating within the gelatinous matrix like a tornado on its side as the slug moves forward.

After the slug has moved to a suitable spot, an even more remarkable transformation takes place. It stands upright, looking like a bowling pin with the cells still whirling around inside. Some of the green cells at the tip—now the top—separate from the rest and spiral downward until they form a sturdy adhesive base. The other green cells form a slender stalk, while the orange cells migrate to the top to form an elegant ball of spores. The final structure has a delicacy that stands in contrast to the slug, like the ugly duckling transformed into a swan.

The purpose of this mighty group effort is to travel a longer distance than any single Dicty could possibly accomplish on its own. Not only is the slug like a cross-country locomotive, but the elevated spores are likely to stick to the body of a passing insect for a transcontinental flight. This group effort requires a degree of coordination that we might never have expected in such a "simple" creature, but it also requires a degree of sacrifice. The cells that form the base and stalk of the structure lose their ability to reproduce. In Darwinian terms they have sacrificed their lives so that other members of their group can reproduce. If a person behaved that way voluntarily in a comparable situation, he or she would be posthumously nominated for the Medal of Honor or sainthood.

From our dispassionate perch on Mount Olympus, we are not here to praise Dicty but merely to wonder how such a division of labor could have evolved, when cheaters who insist upon becoming spores would have a huge advantage over solid citizens who dutifully become stalk and base cells. One possibility is that a given cell doesn't have a choice. Perhaps all that whirling around within the slug is like a tumbler that scrambles lottery tickets, so that the winners (cells that become spores) are based on the luck of the draw. Perhaps all of the cells that stream together to form a slug are genetically identical, derived from a single ancestor many cell divisions ago. Alas, neither of these answers is entirely correct. Slugs frequently contain cells from different lineages, and some of them manage to bias the lottery in their favor. The cellular slime mold is a paragon of microbial virtue, but even Dicty is not without sin.

The conflict that we frame in terms of good and evil exists in all creatures that interact with their neighbors. It doesn't depend upon the complexity or mental activity of the organism, and by now you should be questioning what these terms even mean. I could have regaled you with stories of saints and sinners in lions, elephants, and chimpanzees, but it was unnecessary. The microbes offer hundreds of unseen stories, in and all around us.

18

Groups All the Way Down

ACCORDING TO LEGEND, WILLIAM James was once approached after a lecture by an elderly woman who shared her theory that the earth is supported on the back of a giant turtle. Gently, James asked her what the turtle was standing upon. "A second, far larger turtle!" she replied confidently. "But what does the second turtle stand upon?" James continued, hoping to reveal the absurdity of her argument. The old lady crowed triumphantly, "It's no use, Mr. James—it's turtles all the way down!"

Something strangely similar to an infinite progression of turtles is emerging for our understanding of individuals and groups. Individuals appear much more real to us than groups. Individuals are physically discrete, whereas groups seem like fuzzy abstractions. We endow individuals with mental properties such as intentions and self-interest and regard groups as just a convenient word for what individuals do to each other in their pursuit of self-interest. Actually, in James's day it was more common to think of society as an organism in its own right, but individuals have occupied the center of the intellectual universe for the last half century. British prime minister Margaret Thatcher spoke for her times when she said in a 1978 speech that "there is no such thing as society—only individuals and their families."

As Thatcher was speaking these words, a cell biologist named Lynn Margulis was making the opposite claim—that there is no such thing as individuals, only societies. The cells of all plants and animals (called eukaryotes) are much different and more complex than the cells of bacteria (called prokaryotes). Their DNA is packaged inside a nucleus, and the rest of the cell (the cytoplasm) is inhabited by other complex structures such as mitochondria (which produce energy), chloroplasts (which capture and produce energy from light in plants), and a network of channels called the endoplasmic reticulum. Virtually all biologists assumed that eukaryotic cells had evolved by small mutational steps from prokaryotic cells—individuals from individuals. Lynn claimed that eukaryotic cells evolved from symbiotic associations of bacteria—individuals from *groups*. Many symbiotic associations exist today, such as bacteria and algae that live within the tissues of protozoa. Lynn conjectured that when the members of a symbiotic association become sufficiently dependent upon each other, they make a transition from organisms to organs and the association becomes a new higher-level organism.

The idea was not entirely new. Just as early cartographers noticed that the coastlines of Africa and South America seemed to fit together like puzzle pieces and proposed a theory of continental drift, early cell biologists noticed that mitochondria looked like bacterial cells. Both theories were dismissed as preposterous, however, leaving it to future brave souls to resurrect them. Lynn resurrected the symbiotic theory of the eukaryotic cell against fierce opposition from her colleagues. Even though I have described science as a straightforward process like turning a crank, really big issues such as this one arouse passions more reminiscent of a religious war. Nevertheless, the crank does turn, and today no reasonable person would doubt the evidence for either continental drift or the symbiotic origin of eukaryotic cells. In 1983, Lynn was elected to the U.S. National Academy of Sciences for her achievement, the highest honor that America bestows upon its scientists.

Some aspects of the symbiotic theory remain controversial,

such as the origin of flagella and cilia, the whiplike and hair-like structures that make eukaryotic cells move. Lynn thinks that these are derived from mobile bacteria called spiro-chetes, which evolved to push the symbiotic communities around like tugboats nudging a ship through a harbor. She also thinks that the internal filamentous structures of the cell are derived from spirochetes, including the axons of our nerve cells. Lynn probably wouldn't be happy if she couldn't torment her conservative colleagues with statements such as "All I ask is that we compare human consciousness with spirochete ecology." Even if she proves wrong on some of the details, it remains true that the eukaryotic single-celled or-ganisms of today, which strike us so firmly as individuals, were communities of bacteria in the distant past. A threshold of integration was achieved that caused us to see the whole rather than the parts.

Working upward, multicellular organisms such as you and I are social groups of eukaryotic cells. Working downward, bacteria are social groups of genes. Like an infinite stack of turtles, everything that we recognize as an individual is also a population of subunits. We call the subunits organs, rather than organisms, because they work so well together on behalf of the whole.

The concept of organisms as societies is more than just an arresting metaphor. We have already seen that social life is a contest between "solid citizens" who function well as groups and "sinners" who gain at the expense of solid citizens within groups. The same contest exists among the parts of a single organism. Every time genes replicate within cells and cells di-vide within a multicellular organism, there is potential for some to gain at the expense of others and the welfare of the group as a whole. When this happens, the whole becomes less of an organism and more like a mere group. The subelements become less like organs and more like quarreling organisms with their own separate agendas. The harmony of an organ-ism cannot be taken for granted. It requires the evolution of mechanisms that prevent subversion from within.

Thinking of individuals as social groups has vastly ex-panded the domain of social behavior beyond interactions *among* individuals to the internal organization *of* individuals. Why do chromosomes exist? If genes existed as independent units, they could replicate differentially within the cell like the viruses and bacteria described in the previous chapter. Linking them together into a single structure that replicates as a unit neatly solves the problem of differential replication. Why are the rules of meiosis so elaborate? Meiosis is the process of cell division that results in the formation of sperm and egg cells (gametes). This stage of the life cycle is especially vulnerable to differential replication, so meiosis is elaborately designed to give each gene an equal chance at becoming a gamete. Why is there a division between cells that are destined to reproduce (the germ line) and cells that are destined to build the body of the organism (the somatic line) that takes place very early in development? To minimize the number of mutation-producing cell divisions within the germ line and to ensure that the somatic line is selected exclusively for its ability to enhance the fitness of the whole organism, rather than competing to become gametes. Why are many somatic cells designed to self-destruct after a finite number of divisions? To prevent cheating strategies from evolving over a period of many cell divisions.

These facts about genetics and development are sometimes called laws because they are so widespread, but now they have acquired the other meaning of the word "law"—a social contract designed to promote the common good. My use of words borrowed from human social life, such as "solid citizen" and "sinner," is not just poetic license to indulge a general audience. It is how the experts talk to each other. A single review article titled "The Social Gene" by Harvard biologist David Haig, a recognized authority on the subject, used the following words: allegiance, binding agreements, cabal, cajole, cheat, clique, coalition, coercion, collectives, commons, conspire, contractual arrangement, corrupt, deceit, egalitarian, exploitation, factions, fair play, firm, fraud, free-riders, gangster, huckster, institutions, licensing, lottery, manipulate, marketplace, misappropriate, monopoly, motivation, open society, parliament, partnership, payoffs, police, politics, protection racket, rogue, sabotage, security system, self-interest, social contract, squabbling, steal, strategists, surveillance, swindle, team, trade, transaction costs, and unauthorized. No one fifty years ago could have dreamt that these words would be used to describe genetic, developmental, and physiological interactions within a single individual!

I have shown that large organisms such as you and me are groups of cells, which in turn are groups of bacteria, which in turn are groups of genes. Each level of the hierarchy encounters the inescapable facts of social life: the group-level benefits of working together and the individual-level benefits of exploiting the group. The upper level qualifies as an organism only because the problems that we associate with selfishness and immorality at the lower levels have been substantially solved. But is it really groups *all* the way down? The lowest level that we can go is the origin of life itself.

If the first life forms were individuals, then they were probably naked strands of RNA-like molecules that differed in their sequence of nucleotides (ACGU), the famous four letters of the genetic alphabet (uracil is replaced by thymine in DNA's four-letter alphabet). Each strand (such as AAACCGUU) creates its complementary strand (UUUGGCAA) from nucleotides available in the environment, which in turn recreates the original strand. Different amino acid sequences might vary in their replication efficiency, leading to a primitive form of natural selection. The problem with this scenario—which has been tried in the laboratory—is that RNA replication isn't very efficient in the absence of highly specific enzymes that could not have been present in the primordial soup. The copying fidelity per nucleotide was probably no greater than 99 percent, which might seem high until we realize that a string of 100 nucleotides is almost certain to mutate somewhere along its length at every replication. The result is a broth of very short RNA strands consisting of a single dominant sequence and a halo of closely related mutated sequences, far short of anything that can be called life.

To proceed further, it is necessary to postulate that life began as a community of cooperating molecular reactions. For example, DNA and RNA replication in modern life-forms is made efficient by other molecules that we call replicases, but these helper molecules must also be replicated for the system as a whole to remain intact. As soon as we think about a network of cooperating molecular interactions, we encounter the same inescapable facts of social life that confront higher life-forms. Noncooperating molecular reactions also take place and ultimately destroy the network. As Hungarian biologist Eörs Szathmáry put it in a recent review article, "To provide catalytic support in a molecular catalytic feedback network is an altruistic behavior doomed to exploitation by parasitic molecules and eventual extinction." What we recognize as good and evil in human terms extends even into the realm of chemical reactions!

As with higher life-forms, the (partial) solution to primordial exploitation was probably the existence of many groups of catalytic feedback networks, some of which worked better than others, exactly as in my desert island scenario in Chapter 5 and examples of microbial virtue in Chapter 17. Perhaps the first groups were organized around clay particles or self-organizing lipid vesicles. The origin of life remains a deep mystery, although one that I believe will ultimately yield to scientific understanding. If so, then solving the mystery will not cause it to lose its grandeur. I find it awesome to contemplate that life emerged as tiny molecular fellowships. If so, then like the old woman's theory about an infinite stack of turtles, it's groups all the way down.

19

Divided We Fall

GROUPS BECOME ORGANISMS WHEN selection within groups is suppressed, enabling selection between groups to become the primary evolutionary force. The groups that we call organisms have achieved this imbalance to a remarkable degree. You came into this world as a unique collection of genes that remains the same throughout your lifetime. The process of meiosis makes your sperm or eggs different from each other, but each gene has the same chance of being represented. Meiosis is an equal-opportunity employer. Your stability as a population of genes, combined with the fairness of the rules of meiosis, means that no evolution takes place inside you. Except for mutations, the genes leaving through your gametes are the same as the ones entering through the union of your parents' gametes. The only way that your genes can differentially contribute to the gene pool is by causing you an individual to survive and reproduce better than other individuals in the population.

Well, not quite. Your development from a single cell to a multicellular organism required many cell divisions. In adulthood, your cells continue to divide to replace themselves, especially those that wear out fast, such as your skin cells. Each division is like the tick of a clock in which mutations can occur and spread in competition with other cell lineages. The fact that a successful mutant cell lineage—which we call cancer—might eventually destroy itself along with the rest of the organism is irrelevant. The evolution that takes place in your body has no foresight. It is simply a mechanical process whereby some forms arise and outcompete others on the basis of immediate interactions. As for the fair rules of meiosis, laws are made to be broken. Some mutant genes are able to bias their representation in the gametes above 50 percent, like a passenger muscling his way onto the lifeboat of a sinking ship. The advantage of being overrepresented in the gametes is so great that "meiotic drive" genes, as they are called, can evolve even when they have a debilitating effect on the whole organism. In these and a surprising number of other ways, evolution *does* take place within single organisms, disrupting their harmony and turning them back into mere groups.

To understand cancer, we need to think of a single organism as a vast population of cells, all carrying the same set of genes except for the mutations that are about to become the protagonists of our story. The life of any particular cell is managed from cradle to grave to make sure that it plays its appropriate role. The genes that accomplish this miracle of coordination are given names such as "caretaker genes," "gatekeeper genes," and "landscaper genes," as if the body were a meticulously run English country estate. Even death is managed by a process called apoptosis that disassembles cells in an orderly fashion when they are no longer needed or fail to play according to the rules. Perhaps a totalitarian state is a better metaphor than a British estate.

Cell division is exceptionally well supervised. A new copy of DNA is made, like a monk transcribing a holy text. The copy is proofread and corrected with such accuracy that the final error rate can be less than one in a million for any given letter of the text. Still, the entire text contains millions of letters, so most copies include a few errors. In this fashion, mutations occur with clocklike regularity with every cell division.

Many mutations have no effect on the cell's function. Most of those that do are quickly detected and destroyed by the immine system, which acts like a ruthlessly efficient police force. A knock on the door in the night and the hapless mutants are taken away. However, a small fraction manage to evade the police. These might appear to be enemies of the state or as heroic freedom fighters resisting the tyranny of the state, depending upon whose perspective you wish to take. An even better way to think of them, however, is like prey organisms evolving to avoid their predators. Just as birds remove the most conspicuous insects in the forest, leaving those that blend into their background, the immune system removes the most conspicuous mutant cells in the body, leaving the undetected to survive and grow into tiny populations. There are probably hundreds of these mutant populations in your body right now.

Having adapted to avoid predation, the mutant populations now must compete for resources with the neighboring normal cells that continue to perform their function. In most cases the competition results in a standoff. The mutant populations remain small and of little consequence to the whole organism, neither helpful nor harmful. The mutational clock keeps ticking, however, and additional mutants arise in some of the populations that make them more aggressive in competition with neighboring cells. Just as some plants expand by producing a toxin that they can withstand but their neighbors cannot, the double mutant populations aggressively expand and become incipient tumors.

Now that the incipient tumors have evolved to cope with predators and competitors, new factors limit further growth. Cells are nourished and their wastes are removed by blood vessels. Unless an incipient tumor can recruit blood vessels in the same way as normal tissue, it cannot grow beyond a certain size. Most incipient tumors stop there, but additional mutations occur in some that overcome this limiting factor. This might seem like a large and complicated step, but the genetic instructions for recruiting blood vessels evolved long ago as part of normal development. All the mutation in the incipient tumor must do is activate these instructions, which is relatively simple.

In this fashion, the tumors adapt to the challenges of their environment, mutation by mutation, in just the same way that free-living organisms such as birds and fish adapt to their environments and disease organisms rapidly evolve inside their hosts. The fact that a disease such as the human immuno-deficiency virus (HIV) is a separate organism while a tumor is derived from your own cells makes no difference. In both cases they survive inside your body, do not contribute to your welfare, and tick away like mutational time bombs.

HIV is especially dangerous because it ticks at such a fast rate. Not only does it have a short generation time, but the transcription of its genetic code is like a deliberately sloppy monk dashing off copies of the text without checking for errors. The sloppiness is actually adaptive; despite producing harmful mutations by the dozen, it also produces the next beneficial mutation faster than otherwise. Mutations in tumors have the same effect when they begin to interfere with the exquisite machinery that regulates the lives of normal cells, causing the tumor cells to grow faster and with more errors. Environmental factors that increase mutation rate also have the same effect. In both cases, the mutational time bomb starts to tick faster. The tumor becomes a mosaic of cell populations that differ from each other, creating a new field of competition. The most aggressive clones expand at the expense of other clones, just as the original mutant expanded at the expense of the normal cells.

Just as free-living plants and animals leave areas that are currently occupied to colonize new areas, the final adaptation that ensures the success of a tumor is to disperse, sending forth propagules to colonize new areas of the body. At this point, if not before, the tumor interferes with the vital functions of the organism and the grand evolutionary experiment comes to an end. Tissue invasion and metastasis account for 90 percent of cancer deaths.

Cancer evolution seems like the ultimate in shortsighted-ness. Every mutational "advance" that enables a cell line to avoid its "predators" (the immune system), beat its "competitors" (the normally functioning cells), and colonize other areas of the body (metastasis) brings it closer to its own demise. Even if a tumor remains benign, it will wink out of existence with the natural death of the organism, lacking any mechanism for getting from one body to another. Evolution *within* the organism is incapable of preventing such futility, which is adaptive as far as immediate interactions among the cells are concerned. Only evolution *among* organisms can establish the farsighted outcome of cells that work together for the common good. We are lucky that selection has operated so strongly at this level and for such a long time that cancer is a rare disease that strikes primarily late in life.

Unfortunately, single organisms potentially play the role of cancer cells when they interact in groups of their own. Selection within groups favors shortsighted strategies, and selection among groups is required to establish the farsighted outcome of organisms that work together for the common good, as we have seen. If among-group selection trumps within-group selection as decisively as among-organism selection trumps within-organism selection, the problem appears yet again at the next rung of the biological hierarchy. If we manage to drive ourselves extinct as a species, it will be by the same kind of shortsightedness that causes cancer cells to hasten their own demise.

Winged Minds

THE NEXT STOP ON our journey from the origin of life to human morality and religion is the social insects. The story is much the same—between-group selection trumps within-group selection, turning groups into organisms—but in this case the lower-level units are individual insects and the higher-level units are colonies. Unlike a single organism, a social insect colony is not a physically discrete entity. On any particular day, the workers of a honeybee colony might be flying over an area of several square kilometers. Army ant colonies don't even have a physically discrete home. Astonishingly, these societies of well over a million individuals are permanently on the move. The queen with her swollen abdomen is incapable of walking and is borne on the backs of dutiful workers. The eggs that she lays by the thousands are carried and fed throughout their development by a devoted caste of nurses. When they encounter rough terrain on their travels, some workers spontaneously form into living bridges that the rest of the colony crosses. Social insect colonies qualify as organisms, not because they are physically bounded but because their members coordinate their activities in organlike fashion to perpetuate the whole. The reproductive division of labor between queens and workers is similar to the distinction between germ cells and somatic cells in multi-cellular organisms such as ourselves.

Crossing the threshold from groups *of* organisms to groups *as* organisms is not easy. According to current estimates, this evolutionary leap happened only fifteen times in insect evolution. The advantages of cooperation are so great, however, that these fifteen separate origins gave rise to thousands of species of termites, ants, wasps, and bees, which together account for over half of the biomass of all insects. When you take a walk among woods and fields, most of the prime real estate that you pass is occupied by social insect colonies, with solitary insects filling in the spaces. Ed Wilson describes a social insect colony as a factory inside a fortress, against which solitary insects simply cannot compete. An ant colony moving into a rotten log is like a Wal-Mart moving into your neighborhood: the mom-and-pop businesses must go.

All organisms require a way to coordinate their parts. Physically discrete organisms such as you and I have a system of nerves and hormones. A physically dispersed organism such as a social insect colony must have a comparable system that can operate at a distance. Just as a signal traveling down the axon of one nerve cell must cross the tiny distance of a synapse to enter another nerve cell, it must make a larger leap from one individual to another in a social insect colony. Hormones (defined as chemical signals within the body) must similarly be supplemented by pheromones (defined as chemical signals that travel between bodies). In short, just as the mind of an individual is not contained within any single neuron or hormone, a social insect colony has a mind that is not contained within any single insect.

The concept of a group mind might sound like science fiction, but it has been documented in exquisite detail by social insect biologists such as Thomas Seeley at Cornell University. Tom is the kind of scientist who devotes his life to the study of a single species—honeybees, in his case—and achieves the same kind of laser vision that I described for Doug Emlen and his dung beetles in Chapter 7. I recently visited with Tom on a warm September day in his laboratory, which is located not on Cornell's busy campus but a few miles away in a modest building nestled among fields and forests. The goldenrods and asters were in full bloom, and the silence of the countryside was a delightful contrast to the mechanical din that has become such a constant and burdensome part of our environment.

Tom is a tall man with a monkish appearance. His father was a botany professor at Cornell, and Tom grew up roaming these very fields and forests, a kind of boyhood that is fast disappearing. He remembers becoming fascinated by social insects at an early age when he discovered some beehives next to an old farmhouse. The aroma of wax and honey and the spectacle of thousands of bees crisscrossing the summer sky as he lay in the tall grass left a lasting impression on him. Later, as a high school and college student, he spent his summers working as a helper for a professor of apiculture at Cornell University, at first just painting hives and mowing grass, but eventually helping with experiments and learning the ancient art of beekeeping. Long before science existed as a cultural practice, people have been keeping bees.

In college, Tom pondered whether to make his love of bees a hobby or a profession. His mind was made up by the publication of Ed Wilson's magisterial *The Insect Societies*, which was written four years before *Sociobiology*. Tom was inspired by the opening passage from Ed's book:

> Why do we study these insects? Because, together with man, hummingbirds, and the bristlecone pine, they are among the great achievements of organic evolution. Their social organization—far less than man's because of the feeble intellect and absence of culture, of course, but far greater in respect to cohesion, caste specialization, and individual altruism—is nonpareil. The biologist is invited to consider insect societies because they best exemplify the full sweep of ascending levels of organization, from molecule to society.

Tom was now doubly inspired by his love of bees and the intellectual stimulus provided by Ed. Armed with a degree in chemistry from Dartmouth College and the knowledge of an experienced beekeeper, Tom applied to graduate school at Harvard University to work with Ed and his longtime collaborator on the study of ants, Bert Holldobler. It was the only graduate school to which he applied.

"Wasn't that risky?" I asked during my visit.

"I was a good student," Tom replied without a hint of arrogance.

Smiling, I tried to imagine Ed reading Tom's application and learning about this ace student with a monklike devotion to bees. He was accepted, of course. Just as beekeepers transport their hives to fields and orchards to gather honey and help pollinate the crops, Tom loaded his hives into the back of his pickup truck and drove to Cambridge, Massachusetts, where he set them up on the roof of Harvard's Museum of Comparative Zoology.

That was a quarter century ago, and Tom is now a leading authority on how a honeybee colony makes intelligent decisions as a collective unit. In one experiment he transported a hive into a deep forest with very few flowers for them to visit and provided his own "flowers" in the form of two feeding stations located four hundred meters away on either side of the hive. First one feeder was provided with a more concentrated solution of sugar than the other, and within hours the colony was sending most of its workers to that station. Then the concentrations were reversed, and within hours the colony was sending most of its workers to the other station. How was the colony monitoring the quality of its resources and responding to changes so quickly? Tom knew the answer because he and his assistants had painstakingly marked all four thousand bees in the colony by gluing tiny numbered and colored disks onto their backs, allowing them to be identified as individuals. With an assistant at each feeding station and Tom observing the interior of the hive through a glass panel, the entire decision-making process could be monitored in detail. Most bees visited only one station and therefore had no frame of comparison for making a decision. Instead, when bees returned to the hive from one of the stations and danced to indicate its location to other bees, the duration of their dance was proportional to the concentration of sugar. The other bees did not compare the length of different dances, even though that information was available. Instead they just picked a dancer at random and went to the corresponding station. The fact that some bees danced longer than others created a statistical bias that directed more bees to the better station. The longer-dancing bees attracted more followers simply because they had more time "onstage" than the shorter-dancing bees. No bee made a comparison between the two patches, either by visiting both patches or by comparing the dances of bees that had, but their social interactions enabled the comparison to be made at the colony level.

As another example of decentralized intelligence, the colony acts as if it is hungry when honey reserves become low, sending more workers out into the fields to gather nectar, but the hunger of the colony cannot be traced to the hunger of any individual bee. Instead, when a foraging bee arrives back at the hive, it regurgitates its load of nectar to a second bee that stores it in the honeycombs. When flowers are scarce and honey reserves are low, returning foragers can immediately transfer their load to an idle workforce of storers. When flowers are abundant and most of the combs are full, returning foragers must wait before they can transfer their load to the overworked storers. The amount of time that they must wait provides a reliable signal of the need to gather more nectar. Tom discovered that when foragers can immediately transfer their load (honey reserves low), they are stimulated to dance for a longer duration, which recruits more workers to become foragers. When they must wait to transfer their load (honey reserves high), they dance for a shorter duration and stop foraging themselves. The crucial experiment involved removing a portion of the storing workforce, causing the returning foragers to wait even though honey reserves were low. The colony responded to the false signal by foraging less, as if the combs were full.

These and other examples of decentralized intelligence are described in loving detail in Tom's book *The Wisdom of the Hive: The Social Physiology of Honey Bee Colonies*. The title is derived from a famous book called *The Wisdom of the Body*, published by Walter B. Cannon in the 1930s, which described the amazing physiological processes of single organisms. The wisdom of the hive, like the wisdom of the body, is mind-boggling when understood in detail. A bee colony monitors hundreds of flower patches as they wink in and out of existence over an area of several square kilometers. Foragers must gather pollen and water in addition to nectar and have mechanisms for wisely allocating their efforts in response to changing circumstances. The temperature of the hive is regulated with precision: a bee colony is a warm-blooded animal! Temperature is raised by shivering and lowered by collecting water and spreading it over the surface of the combs, the equivalent of us sweating. Myriad tasks are carried out within the hive, such as building combs, tending the queen and brood, and removing dead workers. All of these functions are carried out by feedback processes similar to what I have described for gathering nectar. There is no centralized intelligence, no single bee guiding the operation, certainly not the queen. It is the pattern of social interactions that creates the wisdom of the hive, just as it is the pattern of neuronal and hormonal interactions that creates the wisdom of a single individual organism.

During my visit, Tom was most eager to describe his newest research on a crucial moment in the life cycle of a colony, when it splits to form a new colony. The queen and about half the workers exit the hive and form into a mass about the size of a football on the branch of a nearby tree, leaving the remaining workers to raise a new queen by feeding one of the larvae a special food called royal jelly. Scouts leave the swarm and fly over the countryside in search of a tree cavity that can serve as a new home. Ten or twenty potential sites are typically found, requiring a decision about which to choose. Tom has been studying whether the swarm is capable of making a wise choice and exactly how it is accomplished. He is excited because the decision-making process is very similar to the interactions among neurons that enable individual organisms to make wise decisions.

Tom explained by describing experiments that neuroscientists have performed on rhesus monkeys that have been trained to watch a collection of dots moving on a screen. Some dots move to the right and others move to the left. The monkeys have been trained to look in the direction that the majority of dots are moving. If they are correct, they are rewarded with a few drops of fruit juice. If they are incorrect, they are punished with a few drops of salt water. The monkeys quickly become proficient at this task, although they still make mistakes when the number of dots moving in each direction is almost equal. While all of this is happening, the scientists are recording the activity of single neurons inside the monkeys' brains, just as Tom observes the activity of single bees inside a colony.

It turns out that some neurons fire only when a dot moves to the right. Other neurons fire only when a dot moves to the left. These two classes of neurons fire at different rates, depending on how many dots are moving in each direction. As soon as the firing rate of one of the classes reaches a certain threshold, the monkey makes its decision and moves its head in the corresponding direction. The monkey's decision is based on a simple competitive process among neurons.

Tom's experiments on honeybee swarms are conducted on a treeless island off the coast of Maine, enabling Tom to completely control the availability of artificial nest cavities that he provides. By altering the properties of the nest cavities, he has determined that scout bees pay attention to no less than seven factors: the volume of the cavity, the size of the entrance hole, the height of the entrance hole, the compass direction of the entrance hole, where the hole enters the cavity (top versus bottom), the presence of combs (colonies frequently starve to death over the winter and their cavities can be reused), and distance from the swarm, with more distant sites preferred over closer sites to avoid competing with the parent colony. The scouts return to the swarm and perform a dance on its surface whose duration is proportional to the quality of the nest cavity, as I have already described for foragers indicating the location and quality of nectar sources. In this fashion, more scouts are recruited to check out the best nest cavities.

It turns out that the final decision is made not at the swarm but at the separate nest cavities. When a scout enters a nest cavity, it does not return immediately but lingers for about an hour. As soon as the number of scouts at a given cavity exceeds a certain threshold, they return to the swarm and perform a new behavior called piping that signals the end of the decision-making process. Tom was especially triumphant about this, his most recent discovery, which he determined by a clever experiment. He placed five identical nest cavities right next to each other, which is too close for the scout bees to distinguish among them. Scouts recruited to the location entered the five cavities at random and therefore accumulated within any one cavity at one-fifth the normal rate. The decision-making process was correspondingly prolonged, proving that the final assessment is made at the nest cavities, not the swarm.

The entire decision-making process takes place among roughly a hundred bees, a tiny fraction of the entire swarm. The other bees (including the queen) remain quiescent, forming the surface upon which the decision-makers dance. As soon as the scouts returning from the chosen site start piping, all the bees warm up their flight muscles, and within sixty seconds the mass expands into a dense cloud about the size of one's living room.

"What happens then?" I asked breathlessly.

Tom gave me a helpless look. He has observed it many times but can't figure out how to study it with the precision that he strives for. He *thinks* that the scouts repeatedly fly through the cloud in the direction of the chosen location, indicating to the others which way to go. Slowly the cloud gathers momentum, like a freight train pulling away from a station, and can move as far as two kilometers before slowing down and stopping in the vicinity of the new site. At this point the scouts gather around the cavity entrance, which can be as small as a knothole, and emit a pheromone that guides the rest of the swarm into their new home.

Even without knowing the details of their final trek, Tom has shown that the process of a bee swarm deciding upon a new home bears an uncanny resemblance to the process of a rhesus monkey deciding which way to turn its head. In both cases, an intelligent decision by the higher-level unit is caused by a mechanical process among lower-level units. The process involves an organized competition among factions representing the alternative choices, whose outcome is based on which faction reaches a threshold first. The intelligence of the upper-level unit cannot be found in any of the parts but rather emerges from the interactions among the parts. Every mind is a group mind, and every mind of a dispersed organism must include social in addition to neuronal interactions.

My meeting with Tom ended because he needed to extract the honey from some of his hives with a beekeeper friend. His friend had found a buyer that would pay a premium price for unfiltered honey for the health food market. "Completely unfiltered, including the body parts of the dead bees," Tom said with a smile. He remains a beekeeper in addition to exploring some of the most fundamental questions that can be asked about the nature of mentality.

21

The Egalitarian Ape

THE LAST FOUR CHAPTERS have covered topics that are fascinating in their own right, but they are also intended as stepping-stones, leading us to the topics of human morality and religion. To review our progress, the conflicts that we frame in terms of good and evil exist for all creatures that interact with their neighbors, even viruses and bacteria. Amazingly, individuals are not just members of groups but groups in their own right. We call them individuals because they have solved the problems of within-group conflict so well. Yet even individuals fall short of the ideal of total harmony associated with the word "organism," as we can see in the case of cancer. Insect colonies raise harmony to a new level, qualifying as organisms even though their members are physically dispersed. A dispersed organism must be coordinated by a "social physiology" and a "group mind" comparable to physiological and neuronal interactions within discrete organisms.

To continue our journey, we need to focus on human evolution. In previous chapters I have compared our species to a mansion with many rooms, most of which are shared by other species. I have described abilities in other species that vastly exceed our own. I have criticized the usual portrayal of human uniqueness as self-congratulatory, as if we alone were intelligent, moral, aesthetic. Nevertheless, we are distinctive in certain respects, especially compared to our immediate primate ancestors. Our capacity for symbolic thought, including but not restricted to language, vastly exceeds that of any other species. So does our ability to socially transmit learned information, which we call culture. Finally, our ability to cooperate invites comparison with multicellular organisms and social insects but far surpasses that of other vertebrates.

"Unique" is the wrong word to describe these special abilities. Evolution requires continuity, so our abilities must have had precursors in the common ancestor that we share with the other living great apes—chimps, bonobos, gorillas, and orangutans—only six million years ago. What were these precursors and what happened to transform them into the three C's of human evolution—cognition, culture, and cooperation?

The probable answer is that we are evolution's newest transition from groups *of* organisms to groups *as* organisms. Our social groups are the primate equivalent of bodies and beehives. Such a transition is a rare event, as we have seen for social insects, but when it happens the consequences are momentous. Mere individuals and disorganized groups are no match for the new group organism, which quickly achieves ecological dominance. The group organism coordinates its mental in addition to its physical activities, as we have also seen for the social insects. In our case, symbolic thought and the social transmission of information are both fundamentally communal activities. The three C's of human evolution are all manifestations of one C—cooperation.

Admittedly, human groups are also different from bodies and beehives in many respects. Members of a human group are not genetically identical to each other, as are the cells of a single organism (apart from mutations). There are no women with vastly distended abdomens, dropping babies by the minute into the arms of celibate nurses. However, these differences do not necessarily matter. Human groups are comparable to bodies and beehives, not because they share any particular feature, but because between-group selection trumps within-group selection in all three cases. The proximate mechanisms that cause this to happen are expected to be different in such different lineages. By the same token, starfish (echinoderms) and sunfish (vertebrates) both qualify as organisms, despite vast differences in their anatomy and physiology.

The human transition bears a closer resemblance to the mechanisms that regulate the social behavior of genes, such as chromosomes and the rules of meiosis. These mechanisms ensure that all members of the group reproduce equally or at least have an equal chance to reproduce, as we have seen. Even highly disparate fates, such as the spore and stalk cells of Dicty the cellular slime mold, can be made fair by a lottery process analogous to drawing straws. The human transition is based on mechanisms that establish equality among members of the group, which is why biologists such as David Haig so easily borrow the language of human social behavior to describe the social behavior of genes, as we saw in Chapter 18.

It might seem strange to describe human social behavior as egalitarian, given the inequalities that exist all around us, but the megasocieties of today are not at all like the tiny social groups that existed during our emergence as a species. For a better comparison, imagine that you are stranded on a tropical island with thirty other people. You realize that you must work together to survive, but there are also temptations to grab what you can for yourself, by force or by stealth. The force option is untenable because no single person is strong enough to bully thirty other people. The stealth option is equally untenable because you are almost never alone. Your opportunities for selfish gain are severely curtailed, if not entirely excluded. Fortunately, everyone else is in the same boat (or, rather, on the same island), so you are unlikely to be cheated, especially if you remain vigilant. The group is egalitarian, not because everyone is virtuous but because they collectively have the means to detect and punish would-be cheaters and bullies.

Television shows are made of this stuff, but the fantasy is a reality for people who live entirely by hunting and gathering and who have resided in small groups since the dawn of humanity. Hunter-gatherers have turned egalitarianism into an art form. Consider the following famous account of a member of the !Kung San tribe (the exclamation point indicates a click that has no counterpart in European languages) from the Kalahari Desert in southern Africa, explaining the customs of his people to anthropologist Richard Lee during the 1970s. "Say that a man has been hunting. He must not come home and announce like a braggart, 'I have killed a big one in the bush!' He must first sit down in silence until I or someone else comes up to his fire and asks 'What did you see today?' He replies quietly, 'Ah, I'm no good for hunting. I saw nothing at all ... maybe just a tiny one.' Then I smile to myself because I now know that he has killed something big."

The jesting continues when they go to retrieve the dead animal: "You mean to say you have dragged us all the way out here to make us cart home your pile of bones? Oh, if I had known it was this thin I wouldn't have come. People, to think I gave up a nice day in the shade for this. At home we may be hungry but at least we have nice cool water to drink."

Lee's informant was perfectly aware of the purpose of all this jesting: "When a young man kills much meat, he comes to think of himself as a chief or a big man, and he thinks of the rest of us as his servants or inferiors. We can't accept this. We refuse one who boasts, for someday his pride will make him kill somebody. So we always speak of his meat as worthless. In this way we cool his heart and make him gentle."

Hundreds of miles away in the steaming jungles of equatorial Africa, anthropologist Colin Turnbull described a similar scene for the Mbuti (popularly known as Pygmies in our culture). After describing the prowess of the best hunters, Turnbull continues: "None of them had the slightest authority over the others. Nor was any moral pressure brought to bear in influencing a decision through personal consideration or respect. The only such moral consideration ever mentioned was that when the band arrived at a decision, it was considered 'good' and that it would 'please the forest.' Anyone not associating himself with the decision was, then, likely to displease the forest, and this was considered 'bad.' Any individual intent on strengthening his own argument might appeal to the forest on grounds that his point of view was 'good' and 'pleasing'; only the ultimate general decision, however, would determine the validity of his claim."

Across the Indian Ocean in the jungles of the Malay Peninsula, the Chewong tribe believes in a system of superstitions called *punen*, which roughly means "a calamity or misfortune, owing to not having satisfied an urgent desire." As anthropologist Signe Howell relates: "In the Chewong world, desires are most likely to occur in connection with food. If someone is not immediately invited to partake of a meal which he observes, or if someone is not given her share of any foodstuff seen to be brought back from the jungle, that person is placed in a state of punen because it is assumed one would always wish to be given a share.... To 'eat alone' is the ultimate bad behavior in Chewong eyes, and there are several myths that testify to this. The sanction on sharing out food originates in the myth about Yinlugen Bud, who was the chief instrument in bringing the Chewong out of their pre-social state by telling them to eat alone was not proper human behavior."

The belief in *punen* includes a ritual that takes place every time food is brought back from the forest. Howell continues: "As soon as a carcass is brought back, and before it has been divided up, someone of the hunter's family touches it with his finger and makes a round touching everyone present in the settlement, each time saying 'punen.' ... This is another way of announcing to everyone present that the food will soon be theirs, and to refrain from desiring it yet awhile. If guests arrive while the hosts are in the middle of a meal, they are immediately asked to partake. If they refuse, saying they have just eaten, they are touched with a finger dipped in the food, while the person touching says 'punen.' "

The calamity that awaits someone placed in a state of *punen* takes the form of an attack by a tiger, a snake, or a poisonous millipede. Since these animals are thought to exist in spirit in addition to material form, almost any misfortune can be interpreted as a consequence of a social transgression.

Humility, myths, rituals, belief in spirits that approve and punish—these traits are unmistakably human and for most people associated with religion, not evolution. Yet their evolutionary significance immediately becomes clear in the context of hunter-gatherer groups, as mechanisms that suppress fitness differences among individuals within groups, enabling the groups to function as adaptive units. They are the human equivalent of chromosomes and the rules of meiosis.

This interpretation is not just idle speculation on my part, based on a few selected examples. It has been elaborated in impressive detail by Christopher Boehm, a man whose vita reads like an intellectual odyssey. The son of a schoolteacher father and librarian mother, Chris majored in philosophy at Antioch College and received his first taste of evolution from a gifted biology professor. He became deeply interested in the nature of morality and obtained his master's and Ph.D. in anthropology from Harvard University. During this period he lived for three years with the Upper Morača tribe of Montenegro (then part of Yugoslavia), leading to his book on conflicts and their resolution in tribal societies titled *Blood Revenge*. As a young assistant professor at Northwestern University, he came under the influence of Donald Campbell, the same man who many years later served as mentor for Kevin Kniffin, whose work with me on beauty was described in Chapter 16. Chris's early interest in evolution was revived by reading Ed Wilson's *Sociobiology* at Don's request. When the Cold War made fieldwork in Yugoslavia impossible, Chris wrote to Jane Goodall requesting permission to study conflict and its resolution in chimpanzees at her field site in Tanzania's Gombe National Park. At the same time, he began to review the anthropological literature to support his argument for the egalitarian nature of hunter-gatherer societies. Few people are better qualified to think about the transition from primate to human society.

As Chris documents in his 1999 book *Hierarchy in the Forest*, egalitarianism pervades not only hunter-gatherer groups but virtually all small-scale human societies. A typical description of the prevailing ethos (from a 1920s account of Alaskan Inuit) is this: "Every man in his eyes has the same rights and the same privileges as every other man in the community. One may be a better hunter, or a more skillful dancer, or have greater control over the spiritual world, but this does not make him more than one member of a group in which all are free and theoretically equal."

The emphasis on equality among *men* is telling. Women

are often (but not always) excluded from the moral circle and dominated by the very men who insist upon equality among themselves. As a member of the Ona tribe, inhabiting the tip of South America, explained to a Westerner who couldn't believe that they lacked a single chief, "We, the Ona, have many chiefs. The men are all captains and all the women are sailors." The moral circle can similarly exclude all members of other groups. When Chris briefly worked among the Navaho Indians of the American Southwest as a graduate student, he was astounded by their lack of aggressiveness toward each other, which did not prevent the Navaho from traditionally making their living by raiding other tribes.

We tend to marvel at this kind of inconsistency as if it were hypocritical, but instead we should marvel at why we marvel. Consistency might be a virtue for a philosopher, but not for an organism, whether the organism is an individual or a group. An organism must adaptively change its behavior depending upon the context, the very opposite of consistency. We expect an individual organism to have a harmonious internal physiology, regardless of whether it acts as a predator, parasite, competitor, or mutualist toward other organisms. Why, then, should we be surprised when a human group such as a Navaho band exhibits internal harmony, even as it rides off to raid other groups? Remember that we are trying to understand how the traits that we associate with morality evolved as biological adaptations, which requires a detached perspective, like the Greek gods looking down from Mount Olympus. Come to think of it, even the Athenians practiced a form of democracy that was restricted to men and excluded women, slaves, and "barbarians," which was their word for anyone who wasn't Greek. What Boehm and others have shown is that egalitarianism is not a cultural invention that began in ancient Greece, as many have supposed, but is part of our genetic endowment that asserts itself whenever appropriate conditions are met. The condition of mutual control is almost invariably met for men in small-scale societies but only sometimes for women. When women are included inside the moral circle, it is probably because they have the means to resist domination by men.

Chris describes small-scale human societies as "moral communities that agree on their values and, as a latent but potent political coalition, are always poised to manipulate or suppress individual deviants." Positive values include (these terms are taken from actual ethnographies compiled by Chris) bravery, competence, drive, generosity, honesty, humility, impartiality, industriousness, modesty, patience, reliability, self-control, tact, trustworthiness, and wisdom. Individual prowess is admired but also feared for leading easily to arrogance, which is one of the cardinal vices. Above all, a member of a small-scale society should not be overbearing and infringe upon the autonomy of others. The combination of communal values and insistence on autonomy sounds strange until we remember the vigilance that is required to prevent the communal virtues from being exploited by members of one's own group.

A small-scale society bristles with defenses against subversion from within. The first line of defense is gossip, which maintains a dossier of information on every member and quickly detects social failings. An example from the Lesu of Micronesia gathered during the 1930s illustrates the power of gossip. One man's pig broke into another man's garden and ate his crop of taro. The offended person was not angry, but the transgressor was alarmed by all the talk that the incident occasioned, which was damaging his reputation. He therefore tried to give the pig to the offended person to "stop the talk." The offended person good-naturedly refused the gift and stopped the talk by declaring that the incident should be forgotten. Gossip functioned like an immune system, detecting and repairing even this tiny breach of etiquette.

When gossip fails, more potent mechanisms are activated. As Chris describes it, "When sanctioning becomes active, a cool manner of greeting can be an ominous signal, one that can quickly lead to direct criticism or even ridicule. Being made fun of by peers is particularly hurtful if one is bound to live forever in a very small cluster of bands and is unable to escape one's reputation. Some potential upstarts are easily put in their place by ridicule, but others may be less sensitive— or more driven. The grievous or habitual offender is likely to meet with harsher sanctions: he may be ostracized by silence or active shunning—or he may be expelled from the group. If an upstart becomes dangerous to the life or liberty of others and is not susceptible to lesser sanctions, fearful or morally outraged foragers go for the ultimate form of social distancing: execution."

The need for such an arsenal of defenses does not imply that everyone is bent on cheating and exploitation. On the contrary, the defense system provides a social environment in which genuine trustworthiness and altruism can thrive, precisely because the wolves of selfishness are being held at bay. We needn't travel to foreign lands to make this point but can prove it for ourselves with a simple experiment. Imagine that you and nine other people are each given $100 to either keep or invest in a group fund. Money contributed to the fund is doubled and distributed equally to every member of the group. If everyone contributes all of their money ($100), each gets $200 in return. However, imagine a cheater who withholds his money while everyone else invests theirs. The $900 contributed to the fund is doubled to $1,800 and distributed equally, so the cheater gets a total of $280 compared to $180 for everyone else—a sizeable incentive to cheat. How much would you invest, and how much do you think the others would contribute? This experiment captures the essence of what I have called the inescapable facts of social life confronting all creatures, from humans to viruses contributing to a "group fund" of gene products available to all viruses within a cell, as we saw in Chapter 17. It has been performed in many different permutations by social scientists, providing a fascinating complement to Chris Boehm's review of actual small-scale societies around the world.

In one version of the experiment, members play the game repeatedly with each other but their contributions are kept anonymous. Most people begin with moderately generous investments but a few yield to the temptation to cheat. Once the others realize they are being exploited, they withhold their own investments and cooperation disappears like water draining from a bathtub. In a second version of the experiment, members are allowed to contribute to a second group fund that is used to punish the cheaters. Specifically, for every dollar contributed to the second fund, three dollars are deducted from the earnings of the cheats. The capacity to punish provided by the second fund completely changes the outcome of the game. When people realize that cheaters are in their midst, rather than withholding their own generosity, they redirect it by contributing to the second fund with zeal. Thanks to the punishers, cheating is no longer profitable and everyone in the group becomes cooperative. What Chris Boehm has painstakingly documented for small-scale societies around the world can be duplicated in the laboratory in a single afternoon.

Righteous indignation is so strong in at least some people that it remains after all material incentives have been removed. In one experiment, subjects observed an unfair social transaction between two other people and were provided an opportunity to punish the transgressor at their own expense. The two other people were completely unaware of the existence of the subject, who therefore had nothing to gain in terms of reputation or future interactions. Nevertheless, a sizeable fraction of the subjects elected to punish the transgressor as the right thing to do.

How does the righteous egalitarianism of small human groups compare to a group of primates? Chris Boehm experienced the difference for himself when, barred from the highlands of Yugoslavia, he found himself toiling through the forests of Gombe National Park in pursuit of chimpanzees. Here is how Chris describes a typical male chimp: "Every young male, as he approaches or reaches adolescence, becomes driven by political aspirations. First, he displays at low-ranking adult females until they begin to pant-grunt submissively when they greet him. Then he moves on to the more formidable females. Sometimes he suffers reverses along the way, particularly if the females have allies to help them. Eventually he will dominate all the females and begin to direct his displays at lower-ranking adult males. If he is successful in that pursuit, he keeps working his way up the male hierarchy until he can go no further."

Dominant male chimps assert their status as relentlessly

as hunter-gatherers keep would-be dominants under their thumb. The chimps stare, bristle, approach menacingly, and outright attack. Subordinates avoid trouble with a crouching posture and fear grin, as if to say, "You're the boss! You're the boss!" Yet chimp society is not completely tyrannical. A male can't achieve high status without allies and often plays the role of peacemaker in conflicts among other members of the group, as primatologist Frans de Waal relates in wonderful detail in books such as *Chimpanzee Politics* and *Good Natured*. In one interaction observed by Chris in the Gombe chimps, a dominant male named Satan put his huge arms around two fighting adolescents and literally pried them apart. Males jointly defend their territory and attack foreign males, sometimes lethally. They cooperatively hunt small game and share what they catch, at least among a select circle of allies. Even the lowest-ranking males get a modicum of respect. In one memorable scene observed by Chris on a mountain slope above Lake Tanganyika, a low-ranking male named Jomeo pirated a baby bush pig that had been captured by baboons. He was immediately surrounded by a gang of more dominant chimps who tore off pieces of meat for themselves, leaving Jomeo "bleating out a thin, continuous, undulating scream and showing all his teeth in a fear grimace." Remarkably, they allowed Jomeo to keep a portion of his prize, even though they could have taken everything for themselves.

Despotic chimp society seems like the polar opposite of egalitarian small-scale human society, yet all that separates them is a shift in the balance of power. Without mutual control, hunter-gatherer society would veer in the direction of chimp society. If we could somehow enable chimp subordinates to defend themselves more effectively, their society would veer in the direction of egalitarianism. If something happened to a chimplike species approximately six million years ago to shift the balance of power in the direction of equality, it would have radically altered the course of its subsequent evolution, compared to other chimplike species, like water cascading down different sides of the Continental Divide.

A molecular biologist named Paul Bingham thinks that he knows what that "something" is. He was not trained to study human evolution, but there is much that outsiders can potentially teach the experts, as we have seen. According to Paul, the key human adaptation is the ability to throw stones. Modern humans are incomparably better at throwing than chimps or any other species, and this ability required many anatomical changes, as the whole-body motion of a baseball pitcher attests. Paul thinks that stone throwing initially evolved to chase away predators and competitors on the African savannah. As soon as we could throw stones with deadly accuracy and force, however, we started stoning each other. An alpha male might be able to intimidate any given rival in hand-to-hand combat, but collectively he can be surrounded and dispatched at a safe distance with stones.

Paul's stone-throwing hypothesis is a specific version of Chris Boehm's more general hypothesis about enforced equality as the key adaptation that makes us so different from other primate species. This, in turn, is a specific example of the transformation from groups *of* organisms to groups *as* organisms that created beehives, bodies, and perhaps even life itself. You might think that knowledge of our origin as a species is forever lost in the mist of time and beyond the reach of scientific inquiry. On the contrary, past events leave clues that can be pieced together with enough ingenuity. That is why we can solve crimes and how we know with near certainty about past events such as ice ages and meteors striking the earth. The key to solving a crime or a scientific puzzle about the past is to have a theory that predicts how the clues fit together better than any other available theory. I do not claim that we have solved the mystery of how we arose as a species, but I think we have made a promising start. From good-natured tolerance to insistence upon conformity, from communal values to jealously guarded autonomy, from profound nonaggression to the cold-blooded exploitation of other groups, from altruistic love to stoning: is there any *other* theory that explains as well the bizarre constellation of traits associated with human morality?

22

Across the Cooperation Divide

IN THE LAST CHAPTER I compared the divergence of our own species from other ape species to water cascading down different sides of the Continental Divide. In our case it was the *cooperation* divide: as soon as egalitarianism became sufficiently established, genetic evolution started to reshape our minds and bodies to function as team players rather than competing against members of our own groups. The rudiments of physical and mental cooperation were always there; it was just a matter of selecting for them rather than against them.

What exactly are the traits that enable us to function as team players? Some of them are extraordinarily simple. Consider our eyes, for example. All mammalian eyes consist of a pupil that admits light into the interior, an iris that acts like the aperture of a camera, and a surrounding sclera that provides a protective outer coat. In our species the sclera is bright white and the iris is colored, providing sharp visual contrast. When we look at someone, we can clearly see where their eyes are pointing independently of where their face is pointing, even at a considerable distance, thanks to the contrast between the sclera and iris. At closer range we can see how much their eyes are dilated, thanks to the contrast between the iris and the pupil. So great is the emotional and communicative power of the eyes that we call them the windows of the soul.

If so, we might be the only primate species that provides windows for others to gaze into. Japanese researchers Hiromi Kobayashi and Shiro Kohshima examined ninety-two primate species and discovered that we are the only one for which the outline of the eye and position of the iris are clearly visible. In every other species the sclera is pigmented to provide low rather than high contrast with the iris and the rest of the face. In addition, the amount of the eye visible to others is disproportionately large and elongated horizontally in humans compared to other primates. Gorillas are bigger than we are, but their amount of exposed eye is actually smaller, giving them their beady-eyed look. With the sole exception of our own species, primate eyes have evolved to be *difficult* to see, to *conceal* rather than *reveal* information about ourselves—the natural equivalent of sunglasses or windows with the curtains drawn.

In Chapter 4, I said that facts are like bricks—humble by themselves but great in combination with other facts. A research grant to study primate eyes would be a prime target for ridicule by American politicians as a waste of the taxpayers' money, but in combination with other facts it becomes a key piece of evidence in one of the greatest unsolved scientific mysteries of our time: the origin of our own species. American scientist Michael Tomasello is at the forefront of unraveling this mystery, although he had to go overseas to do it. He is co-director of the Max Planck Institute for Evolutionary Anthropology in Leipzig, Germany, where he is provided resources that would be difficult or impossible to obtain in the United States.

If you visit the Leipzig Zoo, you will have a rare opportunity to observe all four species of great ape—chimpanzees, bonobos, gorillas, and orangutans—housed separately in large naturalistic compounds. These animals are among the zoo's most popular attractions, but they are also there for another purpose. The compounds have been designed to function as a world-class primate research center in addition to a public display. Not only can Mike and his colleagues observe all four species in their seminatural habitats, but they can also bring individuals into side rooms to conduct carefully controlled experiments. As if that weren't enough, there are laboratories at the Max Planck Institute for conducting behavioral research on human children in parallel with research on apes.

Mike has developed what he calls the cooperative eye hypothesis to explain how our eyes became so different from those of other primates. All apes are acutely aware of other members of their group and pay attention to where they are looking, based on the orientation of the head. However, their use of this information is not necessarily cooperative. In a society where dominant individuals glare at their subordinates, who dare not stare back, natural selection favors the concealment of information. The direction of the head can't be concealed, but the direction of the eyes can by minimizing the amount of exposed eye and the contrast between the iris, the sclera, and the rest of the face. In an egalitarian society it becomes advantageous for members of the team to share information, turning the eyes into organs of communication in addition to organs of vision.

To explore some of these ideas, Mike and his colleagues conducted the same experiment on apes and human children. Individual apes were invited into a side room of their compound with an offer of food, where they sat facing a human researcher on the other side of a Plexiglas panel. After making sure that the ape was paying attention, the researcher performed one of the following actions in random order: (1) lifting the head toward the ceiling with eyes closed, (2) looking up at the ceiling with the eyes while keeping the head stationary, (3) looking to the ceiling with both head and eyes, (4) staring straight ahead with both head and eyes, (5) facing away from the ape and lifting the head toward the ceiling, and (6) facing away from the ape and facing straight ahead. A video camera recorded whether the ape looked up to the ceiling in response to these different actions. The results showed that the apes were over twice as attentive to the direction of the head than to the direction of the eyes.

In contrast, when human children as young as one year old observed the same actions while sitting on their mothers' laps, they followed head movement when the researcher was facing away but relied almost entirely on eye movement when the researcher was facing them. The eyes were the focus of attention, even at such an early age.

An even more remarkable discovery made by Mike and other primatologists concerns pointing. What could be simpler and more natural than pointing at an object to draw attention to it? But this is so only if you are human. Evidently, there is not a single reliable observation of one ape pointing something out for another, using a raised arm or any other equivalent gesture, in captivity or the wild. Apes raised with people learn to point for things that they want but never point to call the attention of their human caretakers to objects of mutual interest, something that human infants start doing around their first birthday.

Apes also don't respond when people try to point things out for them. In one of Mike's experiments, a person shows an ape which of three buckets contains food by tipping the bucket forward so that the ape can see the food. After the ape has become accustomed to this game, the researcher starts pointing to the bucket rather than tipping it forward. Human children easily understand the signal, but apes of all ages are dumbfounded. Yet if the researcher establishes a competitive relationship with the ape and reaches unsuccessfully in the direction of the baited bucket as if to take something himself—a gesture very similar to pointing—the ape instantly understands and reaches for the same container.

Mike explains these mystifying results in one of his papers this way: "If you encounter me on the street and I simply point to the side of the building, the appropriate response would be 'Huh?' But if we both know together that you are searching for your new dentist's office, then the point is immediately meaningful." What comes naturally to humans, even young infants, is a sharing of intentions, an implicit knowledge that we are engaged in a common activity and are supposed to help each other out, if only by calling each other's attention to relevant objects. Our ape relatives evidently lack this awareness, which prevents them from understanding the meaning of pointing no matter how smart they are in other respects.

Mike and his colleagues have performed numerous elegant experiments to demonstrate the sharing of intentions in human children at a very young age. In one, a child is stimulated to point at something to attract the interest of an adult, who responds in one of four ways: (1) simply looking at the object, (2) emoting positively toward the child without looking at the object, (3) showing no reaction, or (4) looking back and forth from the object to the child and emoting positively. Children as young as one year tend to express unhappiness toward the first three responses and are satisfied only by the fourth, suggesting that their purpose was to establish a sharing of interest and attention toward the object. As Mike puts it, "This by itself is rewarding for infants—apparently in a way it is not for any other species on the planet."

In a second experiment, an adult performs an activity such as stapling papers in front of a child without trying to engage the child's attention or participation. Then the adult pretends to misplace the stapler and to look for it. Children as young as one year tend to point toward the stapler, demonstrating shared attention and intention.

In a third experiment, an adult gestures ambiguously toward three toys and says to a young child, "Oh, wow! That's so cool! Can you give it to me?" They have previously played with two of the toys together and the child is familiar with all three. Remarkably, even one-year-old children demonstrate awareness of the adult's previous experience by choosing the toy that is new for the adult.

In a fourth experiment, an adult engages a child in a game that involves passing toys across a table. Occasionally the adult holds up a toy but fails to pass it over, either because she appears unwilling or because she is trying but unable (these pretenses are accomplished in a variety of ways). Children as young as nine months appear to distinguish the intentions of the adult, becoming more upset when the adult appears unwilling rather than unable.

In a fifth experiment, children watch an adult touch her

head to the top of a box to turn on a light. In one condition the adult's hands are occupied holding something, while in a second condition her hands are free. When the children are provided an opportunity to turn on the light, they tend to use their hands in the first condition, demonstrating awareness that the adult had the intention of turning on the light and was using her head only because her hands were occupied. The children tended to use their heads in the second condition, demonstrating awareness that the adult must have had a reason for using her head if her hands were free. These inferences appear as early as eighteen months but not in apes at any age.

These and other experiments demonstrate that human mentality is fundamentally predicated upon sharing. If we don't have shared intention and attention, we can't even do something as simple as point to an object of mutual interest, much less share our behaviors and symbolic representations. Fortunately, sharing has been part of our external social environment for a sufficient period of time that it has become genetically incorporated into our minds, so deeply and subconsciously that we don't recognize it as sharing until we study it scientifically.

Apes that can't point, babies that pick the one toy they know the adult hasn't played with yet—these scientific results amaze us in part because we are so accustomed to thinking of humans as uniquely intelligent and of intelligence as a single thing. It follows that apes should be the next most intelligent creatures on earth, adults should be more intelligent than babies, and something as simple as pointing should be a no-brainer for everyone. All of these expectations are turned topsy-turvy when we think of mental and physical teamwork as the hallmark of human evolution rather than some kind of generic intelligence. The adaptations that make teamwork possible need not be complex, and infants need to partake in them from the very beginning, resolving the paradox of apes that are dumb as far as teamwork is concerned, however brilliant in other respects, and babies that are brilliant with respect to the elements of teamwork, however undeveloped in other respects.

These ideas are so new (largely within the last ten years) that much remains to be discovered. Perhaps they will even fall apart entirely under the continuing gaze of scientific methods. However, they receive additional support from a fascinating and unexpected source—the domestication of dogs.

Dogs are descended from wolves and became domesticated by humans about 135,000 years ago, based on evidence from molecular genetics. Wolf society is different from ape society in many respects but it is still based strongly on dominance. The first wolves to become associated with humans made a transition from a competition-based society to a team-based society, an abrupt version of the more gradual transition made by humans themselves. If mental teamwork depends more on what is favored by natural selection than generic brainpower, then domestic dogs should be better than both wolves and apes at understanding the meaning of pointing and other forms of human communication. This appears to be the case.

In one set of experiments conducted by Hungarian evolutionist Adam Miklosi and colleagues, dogs and human-raised wolves watched a person indicate which of two containers held food by either touching the appropriate container, pointing to it from a distance of five to ten centimeters, or pointing to it from a distance of fifty centimeters. Dogs performed better than wolves, especially when pointing was at the farthest distance, even though the wolves had been raised under doglike conditions. In a second experiment designed to explore why the dogs performed better, the same animals were trained to perform a simple task (such as pulling a string to get some food) and then thwarted in their attempt (by making the string impossible to pull). The wolves and dogs learned the task with equal ease, but when thwarted the dogs looked at their owner sooner and for a longer period of time than the wolves. The dogs regarded their owners as a source of information more than wolves, not because of experiences during their lifetimes but because they are genetically adapted to a social environment where information is freely shared. If you are a dog enthusiast, then you know that the various breeds are capable of feats of communication and coordination that far surpass the demands of these simple experiments. When it comes to mental teamwork, second place goes not to apes but to dogs.

Dogs crossed the cooperation divide by becoming associated with humans a mere 135,000 years ago. Humans have had a much longer time to cascade toward the fertile plains of mental and physical teamwork. Mike concludes his article on why apes don't point by suggesting that the humble act of pointing includes most of the ingredients of language. Both are acts of communication that require joint intention and attention. Both require feedback and an adjustment of the interaction until comprehension is achieved. Both lend themselves to reversals, with the signaled becoming the signaler. Both can be used for a variety of purposes, such as to achieve a desired goal, call attention to something of mutual interest, or help someone else achieve a desired goal. It is easy to imagine a pointing finger leading to the vastly more powerful medium of language, given enough time.

Modern life is a far cry from small-scale society and sometimes becomes as hierarchical and competitive as a chimp troop or a wolf pack. What happens when we cross the cooperation divide the other way? Do we stop talking and pointing things out to each other? Do we avoid eye contact? I wonder if sunglasses are a newfangled way of drawing the curtains back over the windows of the soul, like the opaque eyes of our primate relatives. Are they worn primarily in competitive and hierarchical social environments when they aren't being used to block out the sun? Full of curiosity, I e-mailed Mike about my idea and received the following reply: "I know of no data. But I have seen the World Poker Championship on TV and they all wear sunglasses!"

23

The First Laugh

IT WAS SEPTEMBER 2002 and the start of another academic year. Anne had the demanding task of teaching the introductory biology course to about six hundred students, most of whom were fresh from high school. At five feet two inches and 105 pounds, she was more than a match for them, but it wasn't fun. All of them were bright, but not all were in a state of mind conducive to learning. Some were so bright that they had cruised through high school without trying and only now were discovering the need for discipline. Others had discipline thrust upon them by their families and were discovering the need for self-discipline. Others were highly disciplined but so narrowly concerned with their professional ambitions (the notorious premeds) that they cared only about what it took to get an A. Many were so dazzled or stressed by their social prospects that their courses were an afterthought. As Anne looked out over the sea of faces, only a few appeared genuinely interested, daring to ask a question or coming up after class to explore a particularly interesting point. These were the people who were in a state of mind conducive to learning, and they were not being well served by a large class such as this.

Anne brooded upon their dilemma and decided to do something about it. She announced in class that she would sponsor a one-credit course the following semester for exploring interesting topics in ecology, evolution, and behavior. Exactly what they read and how they organize the discussion would be mostly up to them. The course would not count for any degree requirements and should be taken purely on the basis of interest in the subject matter.

About a dozen students took Anne up on her offer. Most were from her introductory course, but a few heard about it through an invisible grapevine and came from elsewhere. During the next semester they read and discussed books and articles of their choice. Anne was present but did not dominate the discussion. I attended one session but mostly observed Anne's intellectual salon from a distance. What struck me as most remarkable was how these students distinguished themselves during the following years. The best college students go way beyond their narrow degree requirements, piling on additional courses, becoming involved in research, and otherwise functioning as independent thinkers. They are eagerly sought after when they graduate, whatever their career choice might be. These were the students who regarded Anne's offer as an opportunity, even though it didn't advance their career prospects at all in the narrow sense. We will never know whether the experience had a formative effect or merely provided a gathering place for those who were already surging ahead with their interests, but either way it illustrates the importance of a *state of mind* for intellectual development.

One freshman in Anne's class, Matt Gervais, was so self-assured that he asked to take my class for advanced undergraduate and graduate students on evolution and human behavior. As I learned later, Matt had distinguished himself in high school not as a scholar but as an athlete. He was captain of his football team and took first place at the New York State Powerlifting Championships during his junior and senior years. He describes himself as a typical jock, partying and working just hard enough to get good grades, until he saw the movie *The Matrix*, which set his mind aflame with intellectual curiosity. By the time he entered college he was pursuing academic interests with the same drive that he learned on the athletic field.

Matt thrived in my course and had no trouble keeping up with the more advanced students. Evolutionary theory gave him exactly the kind of comprehensive intellectual framework that he was looking for. His term paper was by far the best of the class, as if he had broken away and raced down the field for a touchdown. The assignment for the term paper was to pick any topic to explore from an evolutionary perspective. Somewhat whimsically, Matt chose the topic of laughter.

Laughter, like suicide, homosexuality, adoption, and art, superficially appears difficult to explain from an evolutionary perspective. Why do we do these things when they seem either superfluous or downright detrimental to our survival and reproduction? Yet laughter is anything but superfluous, as I quickly realized upon reading Matt's term paper. Consider some of the basic facts that everyone knows: all people are capable of laughing, except for a tragic few who are severely mentally impaired. Babies begin to laugh between two and four months of age, long before they begin to acquire language. Even congenitally deaf and blind babies laugh, so hearing or seeing the laughter of others is not required. Laughing is primarily a social phenomenon; we know this from personal experience but studies show that we are over thirty times more likely to laugh in the company of others than when alone. A laugh is easily recognizable, despite individual and cultural differences. Laughing is contagious; the mere sound of it makes us want to laugh. Finally, laughing is pleasurable. It puts us in a good mood and makes us act differently than we do in the absence of laughter. If anything qualifies as a genetically innate capacity that requires an evolutionary explanation, it is laughter.

Laughter has been studied and remarked upon for over two thousand years, by Aristotle and Darwin among others, so it might seem that a former football player in his first year of college would have little to offer. If anything, there are too many explanations. Laughter can be invoked by tickling and peekaboo games in babies or by dirty jokes in adults. It erupts spontaneously but also is used strategically. It can be used to invoke good feelings or as a tool of aggression. Do such diverse phenomena even have a single explanation? What Matt attempted in his term paper was similar to what Margie Profet attempted for the subject of pregnancy sickness, as I described in Chapter 12. He tried to fit together the many disparate facts about laughter in the scientific literature like the pieces of a jigsaw puzzle, using evolutionary theory as the picture on the box. Moreover, he brought together pieces from many different scientific disciplines, including primatology, anthropology, psychology, and neurobiology. Evolutionary theory transcends disciplinary boundaries, as Matt had learned from my course and you (I hope) have learned from this book. Matt therefore had two advantages: a good theory to guide his search and more puzzle pieces to work with, compared to experts who remain within their disciplinary boxes.

From primatology, Matt learned that our ape relatives engage in tickling and chasing games accompanied by a facial expression and panting sound very similar to human laughter. Frans de Waal's wonderful photographic portraits of primates in his book *My Family Album* includes one of an adult male bonobo tickling the stomach of an adolescent male, whose mouth is wide open in a big belly laugh unmistakable to the human observer. Thus, human laughter has a clear precursor in apes that occurs during pairwise playful interactions. This is in contrast to human language, which has no obvious precursors in apes.

From neurobiology, Matt learned that there are two different kinds of laughter. Duchenne laughter (named after the pioneering neurophysiologist G.-B.A. Duchenne de Boulogne, whose book *The Mechanisms of Human Facial Expression* was published in 1862) is the spontaneous and emotion-laden response to play in children and humor-invoking situations in adult life such as jokes, puns, and slapstick mishaps, which all share the property of incongruity and unexpectedness in a safe social environment. Tickling, chasing, and the sudden appearance of a face can invoke laughter or fear in children, depending upon how they are interpreted. Slipping on a banana peel is funny unless the person is seriously hurt. Non-Duchenne laughter is more strategic, is less spontaneous, and need not even be accompanied by humor. Normal conversation is often accompanied by laughter, which the speaker uses to highlight a point and the listener uses to signal appreciation. This kind of laughter is smoothly integrated with conversation, in contrast to the more "genuine" Duchenne laughter, which can bring conversation to a gasping, wheezing halt. In addition to lubricating normal conversation, non-Duchenne laughter is used to defuse tense situations (nervous laughter), to make others feel uncomfortable (malicious laughter), and for other strategic purposes. Neurobiological studies show that Duchenne laughter activates ancient brain circuits involved with play and positive emotions in all primates and indeed most mammals. Only non-Duchenne laughter activates areas of the brain associated with our advanced cognitive abilities.

These facts suggest that Duchenne laughter evolved very early in hominid evolution, before we had advanced cognitive abilities. With Duchenne laughter already present in the prehuman behavioral repertoire, it could then be integrated with higher cognitive functions such as language and symbolic thought as they evolved, giving rise to the many functions of non-Duchenne laughter. Simply put, before our ancestors were talking or even thinking in human terms, they were probably laughing.

Matt's synthesis was so compelling that I wondered if it might even warrant publication with more work. Moreover, he fit within a larger vision that I was on the verge of implementing with my colleagues—a campuswide program that uses evolution as a common language to study all subjects for all students and faculty who wish to participate. That vision became EvoS, which I described in Chapter 1 and was initiated the following year. Matt was an inaugural member along with most of the other students from Anne's intellectual salon.

With the help of EvoS, Matt selected his subsequent courses to advance his study of laughter. No single department could contain his interests. He took Anne's animal behavior course to learn more about play in nonhuman species. He took cognitive neuroscience and psycholinguistics courses to learn more about brain mechanisms. He took independent study courses with me to provide even more elbow room. Independent thinkers don't wait to take courses to learn something that they need to know. At any given time, Matt would have several dozen books checked out of the library to pursue this or that line of thought. Like Anne, I acted as a facilitator by providing feedback upon request but otherwise left Matt to his own devices.

Matt's first taste of success came in the form of a Barry M. Goldwater scholarship, which he applied for and received on the strength of his research. This scholarship program is based on a nationwide competition and is highly prestigious, the sort of thing that colleges and hometown papers love to boast about. But could Matt play in the big leagues? During his junior year, we submitted a lengthy manuscript to the journal *Quarterly Review of Biology* titled "The Evolution and Functions of Laughter and Humor: A Synthetic Approach." It was anonymously reviewed by two authorities who knew nothing about Matt's status as an undergraduate student. Their judgment made both of us blush: "An extraordinary synthesis of multiple literatures and perspectives ... careful, nuanced, and persuasive work ... intriguing and accessible from start to finish ... highly influential in years to come ... a gem."

Matt's synthesis is very much in tune with the concept of human groups as dispersed organisms that require social mechanisms of coordination, no less than the social insect colonies described in Chapter 20. The more we began to rely upon each other, the more we needed to feel the same way at the same time. Our brains appear to be wired for this kind of coordination, especially in the form of something called mirror matching systems that become active not only when we do something but also when we perceive others doing it. These systems have precursors in other primates but have been elaborated in our own species to provide what one researcher calls a "shared manifold of intersubjectivity." We can appreciate the movements, sensations, and emotions of others because our brains are literally activated in the same way as if we were doing and feeling them ourselves.

Laughter is an especially effective mechanism for causing members of a group to feel the same way at the same time. Consider an electrical signal traveling down the axon of a nerve cell. When it reaches the end, it causes chemicals to be released into the tiny gap (the synapse) between nerve cells, which in turn cause the adjacent cells to fire, propagating the signal. Now consider someone laughing. If the laughter takes place in an appropriate context, it spontaneously causes others to laugh until the entire group is laughing. The propagation of the signal is almost as fast and automatic as the propagation of an electrical signal from one part of a single brain to another. If the laughter takes place in an inappropriate context (such as a funeral), the laugher will probably receive a baleful look from the rest of the group and will quickly learn from her faux pas. It is difficult to think of ourselves as neurons in a group brain, but that is precisely what we must do to fully understand what it means to be part of a dispersed group organism.

When it is appropriate, laughter puts everyone in a merry mood, as we know from common experience. Mechanistically, the brain releases a cocktail of chemicals similar to those that we take artificially to give ourselves a good time, such as opium and morphine. It is dangerous to take these chemicals artificially because they merely cause us to feel good without causing us to *act* good. Aeons of evolution have caused these chemicals to be released naturally only when they cause us to act in a way that enhances our survival and reproduction. When it comes to brain chemicals, Mother Nature knows best.

Why should a merry mood enhance the survival and reproduction of our ancestors? Life consists of a hierarchy of needs, as humanistic psychologists such as Abraham Maslow fully appreciated. You can't learn new things if you're scrabbling for your next meal or fearing for your life. Our ancestors spent a lot of time scrabbling for food and fearing for their lives as they descended from the trees and ventured out onto the African plains on their two wobbly legs. Periods of safety and satiety were few and far between. Human laughter probably initially evolved as a signal for identifying these periods and quickly establishing a "shared manifold of intersubjectivity" for making the most of them.

The things that our ancestors learned when they were in a merry mood were anything but superficial. Have you ever wondered why a horse or cow grows to adulthood in three years whereas it takes chimps about twelve years and us about eighteen years? The answer in part is that we must acquire more information from members of our group, and our life cycle has been stretched out accordingly. Our prolonged period of juvenile development (compared to other apes) evolved at approximately the same time as our capacity for laughter. The less we relied upon our genes for how to behave, the more we had to rely upon members of our group, which required making the best use of periods of safety and satiety. Fear and hunger are toxic to human development, while laughter is the elixir that makes it possible, for our distant ancestors no less than ourselves.

In three years, Matt developed from a jock dazzled by *The Matrix* to a scientist who made a fundamental contribution to knowledge. This transformation required an exceptional person in a state of mind conducive to learning, but it also required a theory capable of making sense of the world and a way of transmitting the theory that invokes pleasure rather than fear. EvoS was created to provide a supportive social environment for teaching evolution and it is gratifying to see students such as Matt and the others who gathered at Anne's intellectual salon profit so greatly from it.

If I appear boastful, remember that it is about the theory rather than anyone's personal accomplishments. In the long run, it matters little whether I, Anne, Matt, or anyone else is gifted as an individual. What matters is a way of thinking that all capable people can use to understand the world around them. Matt's accomplishment shows how much this way of thinking can add even to a venerable subject such as laughter, which has been pondered by the greatest minds for thousands of years. EvoS shows how this way of thinking can be used to unite an entire university community spanning all subjects. This book is intended to expand the circle still wider. You have learned the same basic principles that enabled Matt to get started. You might not consider yourself a scientist, but you have acquired the background to understand and enjoy the article that we wrote for the experts (available from my Web site: http://www.biology.binghamton. edu/dwilson). Even better, you might become interested in a subject of your own. You might not be able to join us in person, but I can assure you that our conversations are accompanied by humor and laughter.

24

The Vital Arts

WHILE THE RENOWNED HISTORIAN William H. McNeill was a young man working toward his Ph.D. in 1941, he was drafted and sent to Texas for basic training. The only military equipment for the entire battalion was a single inoperative anti-aircraft gun, so most of the training consisted of watching films and marching up and down a dusty field in the sweltering sun to the cadence of "Hut! Hup! Hip! Four!" McNeill didn't need his advanced education to realize that close-order marching had become completely obsolete as far as troop movement and battlefield tactics were concerned. Yet, to his surprise, he discovered that he liked it. As he describes in his 1995 book *Keeping Together in Time: Dance and Drill in Human History*, "Words are inadequate to describe the emotion aroused by the prolonged movement in unison that drilling involved. A sense of pervasive well-being is what I recall; more specifically, a strange sense of personal enlargement; a sort of swelling out, becoming bigger than life, thanks to participation in collective ritual."

Most of us today think of dancing as a mating ritual performed by couples or something that we pay to watch professionals perform onstage, but most dancing throughout history has been a *group* affair. From military drill to ecstatic religious dances, the community dances of villages on festival occasions, and the tribal dances of indigenous people around the world, groups of people assemble to move their bodies in unison, sometimes for so long that they drop from exhaustion or pass into a trancelike state. The effect in all cases is to create a sense of unity among members of the group who have danced together. As the anthropologist A. R. Radcliffe-Brown described for the Andaman Islanders in 1922, "The precise sentiments varied with context, as when two groups danced together after a long period of separation and generated a feeling of harmony, when warriors danced to induce a collective anger before setting out to fight, or at a ceremony of peace-making and reconciliation. On each occasion, the sentiments of unity and concord were intensely felt by every dancer, and this was the primary function of the dance." If we don't want to take a European's word for it, we can listen to the Swazi king Subhuza II, who in 1940 explained how his subjects remained unified: "The warriors dance and sing at the Incwala (an annual festival) so that they do not fight, although they are many and from all parts of the country and proud. When they dance they feel they are one and they can praise each other."

What McNeill discovered by experience as a young man, and returned to explore intellectually later in life, is that the harmony induced by dance is so visceral that it must have a genetic basis. The reason that he couldn't quite describe his euphoria in words is because it emanates from parts of the brain that existed before our capacity for language even evolved. Just as I recounted for laughter in the previous chapter, we were probably dancing before we were talking or even thinking in modern human terms. From a very early stage in our evolution, dancing has probably been part of the "social physiology" that defines and coordinates human groups. Maurice de Saxe, marshal of France during the 1700s, understood the instinctive nature of dance without needing to study it scientifically: "Have them march in cadence. There is the whole secret, and it is the military step of the Romans.... Everyone has seen people dancing all night. But take a man and make him dance for a quarter of an hour without music and see if he can bear it.... Movement to music is natural and automatic. I have often noticed while the drums were beating for the colors, that all the soldiers marched in cadence without intention and without realizing it. Nature and instinct did it for them."

The visceral power of dance made it possible for armies to be formed out of people who had no objective reason for fighting. Merely by marching in time and other intense communal activities, they became emotionally bonded to each other. It is difficult for those who have not experienced this bonding (including myself) to understand its intensity, but J. Glenn Gray puts it this way in *The Warriors: Reflections on Men in Battle:* "Many veterans who are honest with themselves will admit, I believe, that the experience of communal effort in battle, even under the altered conditions of modern war, has been the high point of their lives.... Their 'I' passes insensibly into a 'we,' 'my' becomes 'our,' and individual fate loses its central importance.... I believe that it is nothing less than the assurance of immortality that makes self-sacrifice at these moments so relatively easy.... I may fall, but I do not die, for that which is real in me goes forward and lives on in the comrades for whom I gave up my life."

Even the art of ballet, which might seem like the polar opposite of military drill, was employed as a group-bonding device by Louis XIV of France. At the same time that his inspector general of infantry, Colonel Jean Martinet, was using drill to create the most powerful army in Europe, Louis required the most powerful French noblemen to live at his court for prolonged periods of time, where they took part in elaborate dances and other rituals. A talented dancer himself, Louis took the lead role until the age of twenty-six, after which professional dancers took over, creating the art form of classical ballet. Louis continued to take part in ballroom dances and expected his courtiers to do the same. Dancing unified the nation of France in much the same way as the nation of Swaziland.

As for dance, so also for music, which of course invariably accompanies dance. In his book *The Mating Mind*, my evolutionist colleague Geoffrey Miller attempts to explain music along with all forms of art as sexual display practiced mostly by men for the purpose of attracting women. After all, the typical music video consists of young male singers ostentatiously flashing their wealth amid a bevy of nubile half-naked women. What is there left to explain? Plenty, as Steven Brown, another of my evolutionist colleagues, shows in an article titled "Evolutionary Models of Music: From Sexual Selection to Group Selection." In traditional cultures, music rarely takes the form of young men strutting like peacocks in front of women. Instead, it occurs in many other contexts that coordinate the activities of groups. Perhaps the most comprehensive account of music in a hunter-gatherer society is due to the musicologist Simha Arom for the Aka Pygmies of Central Africa. Their music consists of at least twenty-four discrete categories demarcated by differences in rhythm, scale, instrument type, and performance ensemble. Each category is strictly linked to a social context and performed only in that context—for example, "propitiation for a collective hunt" or "funeral." Each category is instantly recognizable by the Aka and is referred to in their own language by the corresponding social context (e.g., "funeral music"). Steven concludes that "in hunter-gatherer populations like the Aka Pygmies, music is an essential component of literally all social activities, and it does not require a great stretch of the imagination to believe that the same must have been true of our hominid ancestors. . . . At the affective and motivational levels, music is a type of emotive enhancer, and to the extent that music is experienced in a group fashion during ritual events, this enhancement promotes the synchronization of group emotion, motivation, and action."

This interpretation of music is exactly what we should expect when we take the concept of a group organism seriously. If the physiology of a human group is even remotely comparable to that of a body or beehive, it must coordinate many different activities; a general feeling of unity is not enough. Specific musical forms linked to specific social contexts approaches the complexity inherent in the term "physiology."

Modern Western cultures afford many examples of the functionality and specificity of music. Steven begins his article with an account of Dmitri Shostakovich's patriotic Seventh

Symphony, which was composed at the beginning of the Nazi siege of Leningrad: "The Symphony quickly became an important international symbol of the resistance of the Russian people to the Nazi threat. Personal accounts from the period [at]tribute this symphony with giving both the citizens and soldiers of Leningrad the moral strength necessary to resist the siege. So important was the Symphony to the morale of the people that those front-line soldiers who were also trained musicians were transported from the battlefront to the concert hall to perform the Symphony on live radio at the height of the siege on August 13, 1942."

Shostakovich's next symphony was a lament for the death and suffering of his countrymen at Leningrad and elsewhere. Steven continues: "The work is, through and through, an expression of despair and horror. Interestingly, the Eighth Symphony was condemned at the 1948 meeting of the Communist Party Central Committee for its overly catastrophic vision of the war and for Shostakovich's general use of 'formalistic distortions.' " Music *matters*, for us no less than our remote ancestors. Countless examples could be provided from American life, from the protest music of the 1960s to the creepy fear-inducing music of today's political attack ads. The amount of time and money that we spend pouring music into our ears is an indication of how much we require external input to manage our thoughts and emotions. Prior to headphones and iPods, that input came from other members of our group.

As for dance and music, so also for the visual arts. When we think narrowly about survival and reproduction, it appears mystifying that people should spend so much time decorating their implements or producing objects with no utilitarian purpose whatsoever. The primary human adaptation, however, is for our behaviors to be acquired less and less directly from our genes and more and more from other people. The genetic transmission of information is not just a matter of DNA or RNA strands copying themselves. A community of regulated molecular interactions is required, as I showed for the origin of life in Chapter 18. By the same token, the social transmission of information is not just a matter of people wandering around learning what they please; a community of regulated social interactions is required. When we think broadly about survival and reproduction, including the primary human adaptation of social transmission in addition to the daily requirements of survival and reproduction, the question of why we produce art becomes less of a mystery.

Ellen Dissanayake (in books such as *What Is Art For?*) and Kathryn Coe (in her book *The Ancestress Hypothesis: Visual Art as Adaptation*) are pioneering the study of visual art from this cultural evolutionary perspective. Both women have fascinating biographies that include experience with indigenous cultures in addition to the academic study of art. Consider Kathryn's description of how Anna Shaw, a woman from the Pima tribe of Arizona, learned to weave: "Anna spent hours gathering black devil's claw, willow twigs, and cattail reeds with her grandmother as they walked together across the Sonoran Desert. By watching and helping, Anna learned when and how to gather the materials she would need. By helping her grandmother process these materials to get them ready for weaving, Anna developed her own skills. Watching her grandmother weave, Anna saw the basket forms grow and the design motifs emerge. Her young fingers found weaving difficult at first, and baskets had to be woven, then unraveled and rewoven, until the weaving was properly done. She tells us that her grandmother told her 'basket-weaving is to teach you patience, my granddaughter.' When her first well-made basket was completed, she was considered a woman; her adult status began when she gave this basket to her grandmother."

This example shows that the process of *making* art is at least as important as the end product. Kathryn continues: "Those years ... provided an opportunity for a girl to develop a relationship with her grandmother and hear stories about the ancestors they shared, including the meaning of the design motifs they both had inherited from those ancestors, back to the first Pima woman who had woven the first baskets. The ancestors expected the grandmother to teach her granddaughter to weave. They also expected, Anna was told, the granddaughter to learn to weave, to show her generosity by giving her first well-woven basket to her grandmother from whom her weaving skills came.... They also expected her to behave respectfully around the elders; to be loyal and generous; to be a good wife and a good mother; and to cooperate with and be generous to her kin, first her family and then the other Pima. Other Pima were those with whom Anna shared a common ancestor and who could be so identified not only by their language and manner of dress but also by the distinctive baskets woven by the women."

When proficiency at art is required to achieve high status and also requires patience, obedience, respect, and many years in the presence of elders who are transmitting other information in addition to the art skill, the adaptive value of art in the broad sense becomes plausible.

These ideas about dance, music, and the visual arts are admittedly speculative. When I discuss them with my colleagues, I often hear the curmudgeonly slogan "Just-so story!" in response. Speculation is not a sin, however, as I emphasized in Chapter 9. Scientific inquiry always begins with speculation, which is more sympathetically called an untested hypothesis. When we ask *why* these ideas are largely untested, we encounter a problem that goes beyond any particular subject. The very concept of framing and testing scientific hypotheses about dance, music, or the visual arts is foreign to the vast majority of people who love, study, and engage in these activities.

Consider the way that knowledge is organized at institutions of higher education. My university, like most others, includes a College of Arts and Sciences that is divided into a number of divisions. The Division of Science and Mathematics includes physics, chemistry, biology, geology, and one human-related department—psychology. The Division of Social Sciences includes anthropology, economics, geography, history, political science, and sociology. These human-related subjects are often considered "softer," or less scientifically rigorous, than a "hard" science such as physics or biology. The Division of Humanities includes art, English, music, philosophy, religion, and theater. These human-related subject areas are considered not just "softer" than the social sciences but outside the domain of scientific inquiry altogether.

This organization makes no sense from an evolutionary perspective, however deeply entrenched in academic culture. If evolutionary theory is part of the "hard" science of biology, then it can provide an equally rigorous framework for studying our own species. If anything, the subjects associated with the humanities should be *easier* to study from an evolutionary perspective than most subjects associated with the social sciences. After all, dance, music, and the visual arts have all the earmarks of genetically evolved capacities: they appear early in life, are intrinsically enjoyable, exist in all cultures, are mediated by ancient neuronal mechanisms, and often perform vital social functions, as we have seen. By contrast, the "science" of economics is incomprehensible to a toddler, is painful for even college students to learn, exists in only a few cultures, and is mediated by our most recently evolved neuronal mechanisms. It is true that the act of studying economics is more likely to involve scientific methods than the acts of singing, dancing, and creating visual art, but evolutionary theory and scientific methods are required to understand why the latter activities come so naturally and are so vital to our welfare that life does not seem worth living without them.

Unfortunately, that is not how most people who study the humanities for a living see it. They are more likely to regard science in general and evolution in particular as irrelevant to their concerns or even as a threat to everything that they hold dear. Even those who are intrigued by the ideas are likely to be threatened by their lack of basic scientific training, which puts them in the position of a novice rather than an authority. The situation is so bad that it is not yet professionally acceptable for members of humanities departments to flirt with evolution. Bill McNeill is so well established that he can do what he likes. Geoff Miller is a member of a psychology department, where science and evolution are more respectable. Steve Brown supports himself in the medical field of neuroimaging and has given up trying to find a job as a musicologist. Ellen Dissanayake makes a precarious living as an independent scholar without a permanent academic home. Kathryn Coe's day job is in the field of public health. These people are pioneering the study of the humanities from an evolutionary perspective, and not a single one has a job in a humanities department. No wonder that their ideas remain at the speculative stage!

I experienced this kind of resistance for myself when a young man named Jonathan Gottschall appeared at my office door several years ago. He was a graduate student in our English department who wanted to write his thesis on Homer's *Iliad* from an evolutionary perspective but had been forbidden to do so by his major professor. He ultimately prevailed, but only by forming a dissertation committee consisting of members entirely outside his department. After Jon obtained his Ph.D., we decided to publish an edited book titled *The Literary Animal: Evolution and the Nature of Narrative*. When we tried to find a publisher for this volume, I received my own taste of how the academic world can rival religious creationism in its fear and intolerance of evolution.

Jon initially discovered evolution on his own, just like the authors of the *Behavioral and Brain Sciences* articles that I described in Chapter 1. He happened to buy a used copy of *The Naked Ape*, written in 1967 by Desmond Morris, one of the first scientists in recent times to suggest to a popular audience that evolution just might have something to say about our species. At the same time, Jon was also rereading Homer's *Iliad* in a graduate seminar. Here is how he describes his aha experience in the introduction to our edited volume: "As always Homer made my bones flex and ache under the weight of all the terror and beauty of the human condition. But this time around I also experienced the *Iliad* as a drama of naked apes—strutting, preening, fighting, tattooing their chests, and bellowing their power in fierce competition for social dominance, desirable mates, and material resources. Far from diminishing my sympathy for the characters, the revelation of timeless evolutionary logic behind all the petty jealousies, infidelities, spats, lies, cheats, rapes, and homicides that comprise the plot of the poem imparted, for me, a new kind of dignity to it. As Darwin wrote, and as all evolutionists intimately know, there is grandeur in this view of life."

Literary and cultural theorists pride themselves on their tolerance. After all, a major tenet of postmodernism is that all belief systems must be accepted on their own terms. Their tolerance did not extend to evolutionary theory, however, as Jon discovered when he tried to discuss his newfound interest with his professors and peers: "Older professors, like my epics professor, seemed to see 'the naked ape' perspective as a churl's insult to humanity and to great art. The younger professors, as well as my fellow graduate students, saw it as something far worse. I quickly learned that when I spoke of human behavior, psychology, and culture in evolutionary terms, their minds churned through an instant and unconscious process of translation, and they heard 'Hitler,' 'Galton,' 'Spencer,' 'IQ differences,' 'holocaust,' 'racial phrenology,' 'forced sterilization,' 'genetic determinism,' 'Darwinian fundamentalism,' and 'disciplinary imperialism.' ... A close female acquaintance seemed to speak for the whole seminar when she turned to me, shaking her head with a mixture of sadness, pity, and stubborn hope: 'You can't really believe that, Jon, can you?' "

I had several reasons for encouraging Jon and joining forces with him to produce an edited volume. At a personal level, I am the son of a novelist. My father was Sloan Wilson, whose books included *The Man in the Gray Flannel Suit* and *A Summer Place*. I have always felt that my zeal for studying all things human from an evolutionary perspective is similar to the zeal of my father and other novelists for exploring the human condition. At a professional level, literature provides an unparalleled way to learn about other cultures. An epic such as Homer's *Iliad* cannot be taken as literal history, but anyone who wants to know how people thought and acted in a given time and place should read their stories in addition to scholarly reconstructions of their history. Literature reflects basic evolutionary themes that apply to many species, as Jon discovered for the *Iliad*. One contributor to our volume humorously observes that if baboons could read, they would be fascinated by Shakespeare. However, literature also plays an essential role in cultural evolution. Earlier in this chapter I said that the primary human adaptation is for our behaviors to be acquired less and less directly from our genes and more and more from other people. Stories and other narratives surely play a role in this process along with dance, music, and the visual arts.

The Literary Animal was designed to help establish the nascent field of literary Darwinism by identifying and exploring these themes. It would be balanced, including contributors trained in literary analysis who, like Jon, had become interested in evolution, and evolutionists such as myself who were eager to include literature in the study of all things human. It would represent a plurality of views, as it must for such a new enterprise. We were able to assemble an outstanding roster of contributors, including pioneers of literary Darwinism such as Joseph Carroll, biological luminaries such as Ed Wilson, luminaries of literary analysis such as Denis Dutton (who presides over the Web site Arts and Letters Daily in addition to his own work: http://www.aldaily.com), and even the British author Ian McEwan, whose novel *Enduring Love* was inspired in part by evolutionary themes. With such a great subject and cast of authors, I felt certain that we would have our pick of academic publishers.

We first offered the book to the University of Chicago Press, which had just published my book *Darwin's Cathedral*. I enjoyed working with my editor, Christie Henry, so it seemed natural for her handle *The Literary Animal*. The review process went smoothly until, to her amazement, the manuscript was rejected by the executive board of faculty members that oversees the University of Chicago Press. One or more members of the board had reached down like a vengeful god and smitten our book! The University of Chicago Press could publish a book on *religion* from an evolutionary perspective, but it couldn't publish a book on *literature* from an evolutionary perspective!

We then approached other major university presses, including Harvard, Princeton, Yale, Oxford, and Cambridge. In each case the biology editor would express interest but the proposal would get no further than the literary editor. We breathed a huge sigh of relief when we finally found a publisher in Northwestern University Press.

The Literary Animal has just been published as I write these words and is attracting the kind of attention that we hoped for, with reviews and articles in scientific forums such as *Nature* and *Science* and in popular intellectual forums such as the *New York Times Sunday Magazine*. I suppose that Jon and I should feel vindicated and pleased after being turned down by so many publishers, but I look forward to the day when the study of literature, visual art, music, dance, and all other subjects associated with the humanities from an evolutionary perspective is *un*controversial. The current controversy is like the tower of Babel, which cannot be built when everyone is speaking a different language. Evolutionary theory provides a common language that can erase the distinction between the hard sciences, the social sciences, and the humanities. Once we are speaking a common language, then we can proceed to more constructive controversies that result in genuine progress. That is the vision we are pursuing at my university, and I am pleased to report that numerous students and faculty from humanities departments have become enthusiastic participants in EvoS.

Science is often thought to rob the arts of their importance and vitality. How ironic that evolutionary theory leads to a conception of the arts as such an important part of our "social physiology" that they can even be regarded as vital organs.

Dr. Doolittle Was Right

WHO IS NOT CHARMED by the thought of Dr. Doolittle conversing with his parrot, Polynesia, in English and with other animals in their own languages? Of course it is only a story. Language and the capacity for symbolic thought are the crown jewels of human uniqueness, are they not? Perhaps, but in a world where apes don't point and babies can infer the intentions of adults, there is room for a parrot who can think symbolically and talk about it in English.

Our story begins with Terrence Deacon, a polymath whose book *The Symbolic Species* is an intoxicating blend of evolution, neuroscience, and development. The last time I heard him speak, he even wove the formation of snowflakes into his lecture. Terry believes that symbolic thought *is* uniquely human but nevertheless evolved straightforwardly from nonsymbolic thought in apes. This is like pulling a rabbit out of a hat. How can something unique arise from a process that requires continuity? To see how Terry explains it, we must consider the nature of symbolic thought compared to other ways of thinking. Suppose that I repeatedly say the word "cheese" to a mouse while feeding it a bit of cheese. The mouse will quickly associate the word with the object, just as Pavlov's dogs associated the sound of a bell with food. Now suppose that I continue saying the word "cheese" but no longer provide cheese at the same time. The mouse will stop associating the word with the object, as surely as it formed the association in the first place. Now suppose that a perversely rude person says the word "cheese" to you a million times without ever offering you cheese. You might feel like smacking him, but you will still associate the word with the object, even though they aren't paired with each other in your environment. You even associate a word such as "ghost" with something that you have probably never experienced in your entire life!

That's what's special about symbolic thought. With Pavlovian conditioning, the mental associations correspond to the associations that actually exist in the environment. With symbolic thought, the associations become detached from environmental associations, enabling them to acquire a life of their own.

Now for the imaginative flourish that enabled Terry to pull the rabbit of human uniqueness from the hat of evolutionary continuity. Symbolic thought isn't necessarily difficult in computational terms. It might not require a bigger or even a different brain than other species possess. The problem with symbolic thought is that it isn't *useful*. What is the point of creating permanent mental associations that do not exist in the real world when survival and reproduction depend upon figuring out the associations that *do* exist?

The trick to explaining why we became the only species that thinks symbolically is therefore to show how symbolic thought might have become useful for our ancestors, in contrast to virtually all other species. The answer has already been provided in Chapter 22. Symbolic thought is fundamentally communal and requires crossing the cooperation divide. Symbolic thought takes its place alongside our eyes, our ability to point, and our artistic nature as a seed awaiting only the appropriate social environment to grow.

Terry's magic trick has an extraordinary implication: if the rudiments of symbolic thought are not computationally difficult, then perhaps they can be taught to other animals. This would not be a matter of discovering how they already think in their own inscrutable ways, but teaching them to think more like us than their own kind, like a bear riding a bicycle. To see if this has been accomplished, Terry evaluated the numerous attempts to teach our ape relatives to use language. Apes don't have the sound-producing ability to actually talk (something else that evolved in our species along with our communicative eyes), so efforts have focused on teaching them sign language or a system of symbols on a computer console that can be pressed in sequence, just as words are uttered in sequence. The computer method is used by Sue Savage-Rumbaugh and Duane Rumbaugh at Georgia State University and has the advantage of recording every keystroke, providing a detailed record of the learning process and the resulting system of thought. For example, some keys refer to verbs such as "eat" or "drink," while other keys refer to nouns such as "banana" or "juice." Correct sequences such as "eat banana" and "drink juice" are rewarded, while incorrect sentences such as "drink banana" or "juice eat" are not. Apes become adept at learning the right combinations, although some are much better than others and many hundreds of trials are required even for those that succeed. The question is, have they merely learned a large number of specific associations, or have they begun to understand a more general system of relationships represented by the symbols? The test is to introduce a new symbol referring to a new object, such as "grape." If the ape is thinking symbolically, it will realize that a grape falls within the category of "solid food"' and will immediately say (by pressing the keys in the appropriate sequence) "eat grape." There is nothing to learn because a grape falls within a system of categories and relationships that has already been learned—that's what symbolic thought is all about. If the ape has merely learned a large number of specific associations, then it will tediously say things like "drink grape" and "grape banana" before learning which combination is rewarded. According to Terry, only some ape language experiments have succeeded in teaching symbolic thought, and these required very specialized training procedures. Even if it only happened once, however, it would demonstrate the extraordinary possibility that a nonhuman species *can* be taught to think symbolically.

By far the champion symbolic thinker among apes is a bonobo named Kanzi. As a baby, Kanzi was raised by a chimpanzee foster mother named Matata. It was Matata who was the subject of a language-learning experiment, with baby Kanzi crawling around and making a nuisance of himself in the background. Yet Kanzi far surpassed his mother without any of the formal training that she was receiving, just as human infants learn language spontaneously while adults learn a second language only with difficulty. Kanzi became so good that he learned to understand English even though he could speak only through the symbols on the computer console.

Kanzi has been featured in many science documentaries, including an episode of the Public Broadcasting Corporation's *Nature* series titled "Monkey in the Mirror." The sequence begins with Kanzi on a picnic with two of his human caretakers, expertly slicing an apple with a knife to wrap in tinfoil and place on the fire. "Close 'em up!" one of the caretakers tells Kanzi, and the bonobo folds the foil around the apples. To the request "Can you put the apples on the fire?" he nonchalantly tosses the foil package onto the frying pan. Back in the laboratory, Kanzi is shown sitting among an assortment of objects with his caretaker in front of him wearing a welder's mask to conceal any inadvertent signals that might be conveyed through facial expressions. "Could you put some soap on the ball?" she asks, and Kanzi responds by squirting some liquid soap onto a basketball, even though he has never been asked to do this before. Kanzi continues to respond correctly to a series of increasingly bizarre requests, such as "Can you put the pine needles in the refrigerator?" When he is asked to pour the Perrier water into the jelly, he correctly chooses the bottle of Perrier water rather than the jug of regular water. When he is asked to get the pine needles that are in the refrigerator, he ignores the pine needles that are by his side and retrieves the ones that he previously placed in the refrigerator. The sequence ends with the most bizarre request of all: "Take the vacuum cleaner outdoors." Kanzi begins by gathering up the electric cord of a shop vacuum cleaner but then stops and looks at his masked partner as if to say "You want me to do *what*?" She repeats the request, and he pulls the

vacuum cleaner by its cord to the door of his outdoor enclosure, picks it up, and places it through the door. These and
other demonstrations provide compelling evidence that Kanzi
understands spoken English and has gone far beyond learning
specific associations.

Parrots have the vocal equipment to speak human language. Might *they* be taught to think symbolically, enabling
them to actually talk to us like Dr. Doolittle's parrot? This
possibility is not as ludicrous as it might initially sound. In the
first place, intelligence is not a linear scale with humans at
one end, apes next, and something like a parrot far down the
line. Even evolutionists make this mistake again and again.
We swoon with delight when a chimp uses a stick to probe
for termites or a gorilla picks up a branch to lean upon when
crossing a stream, as if we are seeing the dawn of tool use in
our own species. Then we rub our eyes in disbelief when
crows from the Pacific island of New Caledonia (but not elsewhere) fashion hooks to snag beetle grubs out of the nooks
and crannies of trees. It turns out that many parrot and corvid
(the family that includes crows) species have the same kind
of flexible intelligence that we associate with primates, in
part because they live in the same kind of social groups and
forage for the same diverse array of foods. In the second
place, we are not necessarily assuming that parrots evolved to
think symbolically, any more than apes. The question is
whether they can be trained to think more like us than their
own kind. In the third place, if Terry Deacon is correct, symbolic thought isn't necessarily that complicated, as long as
one knows what to train for.

It took an outsider named Irene Pepperberg to attempt
something as outrageous as teaching a parrot to truly talk.
Irene was an only child who grew up in an apartment above a
store in New York City, where she spent a lot of time talking
to her pet parakeet. Her father wanted to be a biochemist but
became a schoolteacher when he couldn't afford college. Her
mother was a homemaker who occasionally worked as a
bookkeeper. Irene was initially afraid of chemistry but, encouraged by her father, discovered that she was good at it.
She graduated from MIT and was working on her Ph.D. in

theoretical chemistry at Harvard when she started watching the science and nature programs such as *Nova* that were appearing on public television. Until then, her only exposure to biology at college was the "slice-and-dice" variety (as she puts it), and the only nature program on TV was *Wild Kingdom*, featuring Marlin Perkins and his beefy sidekick Stan Brock, wrestling anacondas and performing other manly feats.

The idea that you could make a career out of studying animals without slicing them up was a revelation to Irene. Harvard was a center for this kind of research, but it was invisible to someone studying theoretical chemistry. When she saw programs on the ape language-learning experiments, the idea of using parrots occurred to her immediately. It was too late to change her formal education, so she spent her last two years at Harvard sitting in on the courses that she wanted to take while continuing to earn her chemistry Ph.D. Upon graduation, she followed her husband to Purdue University, where he had a job and she was provided an office, just like my wife, Anne. It was there that she purchased Alex, an African grey parrot that would make both of them famous. The first grant proposal that Irene wrote to fund her research was rejected. As Irene puts it, the tenor of the reviews was "What are you smoking?" Ultimately she obtained funding and established her own career, but only thanks to supporters who regarded her work as revolutionary and were willing to fight for her.

Irene trained Alex by including him in a three-way conversation with another person. For example, Irene might hold up a cork and ask the other person, "What's this?" To the reply "Cork," Irene would hand the cork to the other person, who would delightedly shred it to bits. Then they would turn to Alex and ask him what it was. If he replied with a sound close to "cork," he would be given the cork to shred. The object was the reward, rather than something else such as food for giving a correct response. Only later, when Alex had built up a vocabulary and could ask for something else as a result of giving a correct response, was he rewarded with food.

Many years later, Alex displays the same overwhelming evidence for symbolic thought as Kanzi, made even more amazing by being expressed in spoken English rather than keystrokes on a computer console. In one training session shown in an episode of *Scientific American Frontiers*, Alex is shown identifying the shapes and colors of various objects but demonstrates true intelligence when he loses interest and declares, "I'm going to go away!" In addition to his verbal ability, Alex can count the number of objects in front of him up to six and has mastered the concepts of bigger/smaller, same/different, and the absence of difference. For example, when asked about the difference between red and blue blocks of the same size, he replies, "Color." When asked about the difference between two blue blocks of the same size, he replies, "None." He even knows that one number (such as six) is greater than another (such as four) when shown the Arabic symbols!

Alex's finest moment came during a demonstration that Irene was giving to some wealthy potential benefactors of her research. She was showing that Alex had learned the sound of each letter, but Alex was primarily interested in eating nuts. Holding a tray of colorful refrigerator magnet letters in front of him, she asked, "What sound is blue?"

Alex correctly pronounced the sound of the blue *T* magnet and added, "Want a nut."

Irene tried to fend him off by asking another question: "What sound is purple?"

Alex correctly pronounced the sound of the purple *S* magnet and repeated, "Want a nut!"

Irene tried to ask another question, but Alex lost patience and cut her off with "Want a nut! N-U-T," sounding out all three letters of the word!

It is important not to exaggerate the accomplishments of Alex, Kanzi, and other animals that have been taught to think and communicate symbolically. It's not as if they have intimate conversations with their caretakers. Over 90 percent of Kanzi's keyboard utterances are about getting things for himself. Alex similarly speaks up primarily to get food or attention. It is also important to be modest about what has been demonstrated scientifically. Perhaps these animals have been taught to think more like us than their own kind, like the bear riding the bicycle, or perhaps they think and talk symbolically to each other in languages that we have yet to decipher. It is amazing that they can be trained to think and talk symbolically at all, but perhaps even better training methods can be devised in the future. We will never know unless we continue this kind of research, which unfortunately has become as precarious in the United States as the research on silver foxes described in Chapter 7 has become in Russia. Irene's funding from federal sources such as the National Science Foundation ended in 1999, forcing her to search frantically for private donors and even sell Alex mugs and earrings made from his feathers. Visit the Alex Foundation (http://www.alex foundation.org) on the Internet if you wish to make a donation. Alex and two other parrots in training, Griffin and Arthur, are forced to share a single room, making it difficult to train them independently and conduct controlled experiments. The crowded conditions do enable Alex to play the role of a curmudgeonly teacher, however. When the younger birds are being trained, he occasionally chips in with comments such as "Talk clearly!"

Despite all that is not known, Alex and Kanzi tell us something tremendously important about human evolution. We need not postulate a magical mental creation event to explain the crown jewels of human uniqueness. The seeds of symbolic thought lie latent in apes and even parrots, awaiting the appropriate environment to be favored by natural selection, the fertile environment that we gained access to by crossing the cooperation divide.

How Many Inventors Does It Take to Make a Lightbulb?

HOW MANY QUALITY MANAGERS does it take to change a lightbulb? We've formed a quality circle to study the problem of why lightbulbs burn out and determine the best thing we as managers can do to enable lightbulbs to work smarter, not harder.

Lightbulb jokes (there are thousands of them) are funny because they portray groups of people doing badly what a single person can do with ease. Oddly, much of the scientific literature on how people think in groups reads like a lightbulb joke. Take the concept of groupthink, a term coined by Irving Janis in the 1970s to describe groups as dysfunctional decision-making units. As he put it: "I use the term 'groupthink' as a quick and easy way to refer to a mode of thinking that people engage in when they are deeply involved in a cohesive in-group, when the members' strivings for unanimity override their motivation to realistically appraise alternative courses of action. Groupthink refers to a deterioration of mental efficiency, reality testing and moral judgment that results from in-group pressures." Janis interpreted a number of famous foreign-policy disasters as examples of groupthink. Evidently, crafting a foreign policy is like changing a lightbulb—something best done by individuals and botched by groups.

Or take the concept of brainstorming. In 1939, an advertising executive named Alex Osborne claimed that his employees generated more and better ideas in groups than alone. Psychologists conducted dozens of experiments to test this claim by comparing the performance of brainstorming groups with the performance of an equal number of individuals thinking by themselves (called nominal groups). The results were so uniformly negative that a 1991 review article concluded: "It appears to be particularly difficult to justify brainstorming techniques in terms of any performance outcomes, and the long-lived popularity of brainstorming techniques is unequivocally and substantively misguided."

These seemingly authoritative results challenge the concept of mental teamwork that I have sketched in the last five chapters. I would have expected just the opposite. Honeybees think in groups so well that they merge into a single mind, as I showed in Chapter 20. If members of human groups merge their muscles in cooperative physical activities, shouldn't they merge their minds as well? We have been living in groups for our entire history as a species and the higher the stakes— the more "deeply involved in a cohesive in-group"—the *better* our collective decision making should become. The idea of human evolution as one long lightbulb joke just doesn't make any sense—yet it seems to be supported by an enormous body of scientific research.

I decided to resolve this issue for myself about ten years ago. You might be interested in how I approach a new subject such as this without any prior training. My first step was to type the keywords "group decision making" and "group problem solving" into PsychInfo, a computerized service specialized for searching the psychological literature. With the click of a button, I received the citations and abstracts of 495 articles, which I printed and placed in a large three-ring notebook. It would take a long time to read that many articles, but the abstracts can be read in just a few hours. I did it over a weekend, making an index of the most relevant articles that seemed worth reading in full as I went along. As expected for the Ivory Archipelago, the articles represented many different research groups that only partially communicated with each other. Right away, I had a bird's-eye view of the subject that surpassed most of the experts. As my inquiry deepened, I would meet or e-mail some of the authors and ask, "Are you familiar with so-and-so's research on such-and-such?" Often they were not, placing me in the unusual position of advising them on their own subject.

I now had a box full of puzzle pieces to assemble, using evolutionary theory as the picture on the box, just like Margie Profet for the subject of pregnancy sickness and Matt Gervais for the subject of laughter. In addition, I wanted to conduct my own research. I had an idea that the mental tasks used in most psychological experiments were not sufficiently difficult. After all, a trivial physical task such as changing a lightbulb *is* best performed by a single individual, and more people only get in the way. It takes a more difficult task, such as moving a piano, to demonstrate the advantages of physical cooperation. Perhaps I could demonstrate the advantages of mental cooperation where previous efforts had failed by making the mental task sufficiently difficult.

Throughout this book I have tried to portray science as a down-to-earth, roll-up-your-sleeves activity, like gardening or construction work. It's hard work, but not *too* hard because the rewards are sufficiently great. So much effort is required because science is all about accountability. I needed to *prove* that groups surpass individuals for at least some mental tasks to a community of very smart people who were likely to be very skeptical of my claim. My methods and results would be scrutinized at a very fine level of detail. One false move might cause my study to be rejected for publication or discredited after publication.

A job this hard requires help, no matter what psychologists might say about groupthink and brainstorming. My partners for this project were Ralph Miller, a distinguished psychologist at my university, and my graduate student John Timmel. Ralph conducts meticulous experiments on learning in rats and humans but also enjoys accompanying me on my barbaric incursions into his field. John is the son of a corporate lawyer father and homemaker mother who grew up on Staten Island, next to a park with scrubby woods and marshes. He loved collecting snakes and frogs, and his mother indulged him by allowing him to keep a menagerie of pets. He attended Catholic schools and enlisted in the Navy so that he could attend college on the G.I. Bill. He entered graduate school thinking that he would study something like snakes and frogs but quickly became fascinated with human behavior from an evolutionary perspective after taking one of my courses.

Our first challenge was to decide upon the mental task that would be performed by individuals and groups. After much thought, we chose the children's game of Twenty Questions, which involves guessing a word by asking questions that can be answered with "yes," "no," or "ambiguous." We chose this game for a number of reasons. It is familiar to most people, so lengthy instructions are not required. It is a difficult mental task, especially when the word to be guessed is obscure. For any given game, the difficulty increases as each question and its answer adds to the load of information that must be kept in mind to ask the next question. Finally, the game can be played by either individuals or groups. How well would groups perform compared to individuals, and would their advantage increase with the difficulty of the game?

The word that we had them guess was a job title, defined as anything that can be done for a living. To obtain a list of job titles, prior to the experiment we had forty students each list as many job titles as they could think of during a one-hour period and compiled their responses into a master list of 442 job titles. We also kept track of the number of lists on which a given job title appeared. For example, "doctor" was included on all forty lists, whereas "bricklayer" was included on only one list. This by itself says something interesting about the organization of memory. Everyone knows that "bricklayer" is a job title, but almost no one *recalls* it when asked to list job titles! The number of lists on which a given job title appeared provided us with an index of recall difficulty. If mental cooperation is required primarily for *difficult* tasks, then the performance advantage of groups should be greater for words such as "bricklayer" than for words such as "doctor."

Next we needed to round up some people to play the game. Most psychology experiments are performed on college students who are taking an introductory psychology course and are required to act as guinea pigs for thirty minutes or an hour in several different experiments over the course of the semester. We needed the students for a longer period of time, so we created our own full-length course titled "The Evolution and Psychology of Decision Making." The experiment would take place during the first few weeks of the course with the students acting as guinea pigs. Then they would help to analyze the experiment during the rest of the course along with reading and discussing the scientific literature. It might seem strange to use a course to perform an experiment in this way, but it provided a much richer educational experience than your average course, as the students themselves emphasized in their final evaluations.

After weeks of planning, our grand experiment finally began when thirty-six students filed into class at the start of the semester. Half played the game as individuals and the other half played as groups of three individuals for five one-hour sessions over a two-week period (Phase I). Then their roles were reversed for an additional four one-hour sessions (Phase II), so each student played the game as an individual and as a member of a group. Each session took place in a room without any furnishings other than a table and chairs to avoid cueing effects. The word to be guessed was chosen at random from the master list, and as soon as one game was finished another began until the hour was up. There was no time limit for playing a game, and partially completed games at the end of the hour were discarded from the analysis. Nothing could be written down, but individuals and groups were otherwise free to play the game however they liked. The monitor (either John or an undergraduate assistant who was not a student in the course) wrote down each question, answer, and the time required to ask each question, providing a record of each game. Students were instructed not to talk about the games and especially not to discuss strategy outside of the one-hour sessions.

It might seem tedious to list all of these details, but that is the price of scientific accountability. Ralph, John, and I were anything but bored as we discussed the pros and cons of whether the groups should be same-sex or mixed-sex, the need for John and his assistant to conduct themselves as monitors in the same manner, and dozens of other points to make sure that *everything* is held constant except for the planned comparison of individuals versus groups. For example, suppose that we had decided to have everyone play the game first as an individual and then as a member of a group. That would be a fatal flaw because our results could be attributed *either* to a group effect (individuals versus groups) *or* to a sequence effect (early versus late). The ocean floor of science is littered with the wrecks of experiments that sank on the basis of fatal flaws such as these, never to be heard from again. We wanted to avoid that fate for our experiment, which gave us an intense interest in the details.

As soon as the experiment was over, Ralph, John, and I scrambled to analyze the results to report to the students. Overall, groups solved twice as many games as individuals. Their performance advantage was greater for obscure job titles (such as "bricklayer") than for common job titles (such as "doctor"), exactly as we predicted on the basis of task difficulty. Surprisingly, there was no tendency for individuals or groups to get better over the five sessions of Phase I or the four sessions of Phase II. Moreover, the students who played in groups during Phase I did not perform better as individuals in Phase II. Evidently, the groups did not learn certain strategies that could then be employed by individuals. Instead, the advantages of playing as a group required *being* in a group.

We then compared the performance of individuals when playing by themselves to their performance as members of a group. Some students were much better solo players than others. One of the best solo players went on to earn his MBA at Yale University and became an investment banker. When this person joined a group he quickly became their leader and instructed them how to play the game, yet that group performed poorly compared to other groups whose members fared poorly as individuals! Overall, there was no correlation whatsoever between the performance of individuals playing solo and their performance as members of groups.

These results began to make sense when we reflected upon

the actual behaviors displayed by individuals and groups during the sessions. Individuals began by asking sensible questions such as "Is a college degree required?" or "Is the job performed indoors?" but as the game proceeded they began to struggle under the weight of the information that they needed to keep in mind to ask the next question. Some described the experience as "agonizing" and dreaded playing hour after hour by themselves. Groups, in contrast, tended to burst into spontaneous conversation. A party atmosphere often developed, with cheers for questions that turned out to be correct and groans for questions that seemed to go nowhere. The group organized by the future MBA might have performed poorly because the other members followed his instructions without much enthusiasm and a spontaneous atmosphere failed to develop. The average group not only performed better than the average individual but also had more fun!

These results supported my original hunch that a *challenging* task is required to demonstrate the advantages of mental cooperation, no less than physical cooperation. In the game of Twenty Questions, asking the first question is like changing a lightbulb, but asking the tenth question is like moving a piano. Moreover, the spontaneity of group performance is exactly what one should expect from an evolutionary perspective. Thinking in groups isn't something that we must learn to do, like mathematics, but something that we evolved to do without learning, like seeing. We open our eyes and see without any instruction, even though vision requires mental processing that is mind-bogglingly complex when studied scientifically. We burst into enjoyable spontaneous conversation about matters of mutual interest without any instruction, and the mental processing that takes place among individuals might be similarly complex. It takes an ounce of evolutionary thinking to even imagine group decision making as like vision, and our humble experiment suggested that we were on the right track.

To proceed further, we embarked upon an analysis that required the participation of the whole class after the experiment was over. Consider a game of Twenty Questions in which the word to be guessed is "sailor" and the first question asked is "Does the job require a college degree?" This question can be answered not only for the job of sailor but for all of the jobs on the master list, allowing the fraction of jobs excluded by the first question to be determined. The second question, such as "Is the job performed outdoors?" can similarly be answered for all the job titles on the master list that were not excluded by the first question. In this fashion, the sequence of questions in a single game can be represented as a declining curve for the number of job titles on the master list remaining for consideration. The declining curve is steep for a well-played game and shallow for a poorly played game, regardless of whether the job title is actually guessed by the end of the game. This analysis required the construction of a large file in which all of the questions asked by either individuals or groups for a number of games—more than eight hundred questions—were answered for all 442 job titles on the master list. If our experiment was like building a house, this analysis was like a community barn-raising event. It was hard work, but not *too* hard because the results were very gratifying indeed. The declining curves were identical for individuals and groups for the first few questions of the games but then became markedly steeper for the groups compared to the individuals. Groups surpassed individuals as decision-making units only when the mental going got rough.

Scientists are a skeptical lot. Even when an experiment seems airtight, they still often reserve judgment until another laboratory performs the same experiment and gets the same result, or the same laboratory demonstrates the same result using different methods. After the class was over, Ralph, John, and I therefore conducted a second experiment in a brainstorming format to confirm our results. As I mentioned at the beginning of this chapter, brainstorming experiments compare the number and creativity of ideas generated by real groups to so-called nominal groups whose members generate ideas on their own. The ideas of the nominal groups' members are combined and duplicates are removed (much as we created our master list of job titles) to compare to the real groups. The typical result is that nominal groups are no less creative and almost invariably surpass real groups in the number of ideas. To our surprise, however, none of the many brainstorming experiments had examined the importance of task difficulty. We therefore compared real and nominal groups for an easy and a difficult task. The easy task was to think of as many job titles as possible. Even this task is not so easy, since it is hard to recall obscure job titles such as "bricklayer," as I have already mentioned. The difficult task was to think of as many job titles as possible that satisfy the answers given to seven questions, as in a game of Twenty Questions (we used an actual game from our first experiment). This experiment required only an hour for each participant and was therefore much simpler to conduct than our gargantuan first experiment. Real and nominal groups performed equally well for the simple task, but real groups generated 50 percent more job titles than the nominal groups for the difficult task. Amazingly, the entire scientific literature on brainstorming might have reached a false conclusion about the advantages of thinking in groups by confining itself to mental tasks comparable to changing a lightbulb, as opposed to moving a piano.

As these experiments were in progress, I was also trying to assemble the 495 abstracts in my three-ring binder, like so many puzzle pieces, into a single coherent picture. Each abstract described research that involved roughly the same amount of effort as our study. Each had survived the scientific review process to become published and therefore documented facts that were likely to be true. Yet these facts had been assembled into a larger picture of groups as *dysfunctional* at mental tasks, which made no sense from an evolutionary perspective. The more familiar I became with these articles by reading them in full, the more support I found for the *advantages* of thinking in groups. The problem was not with the facts but with the way they had been assembled to reach a larger overall conclusion.

Some of the articles had the gemlike quality of a good short story. My favorite was an experiment by Arie Kruglanski and Donna Webster on Israeli Scout troops deciding between two sites for a work camp, one objectively better than the other. Individuals had previously rated each other for liking, appreciation, and respect, which were combined into a single index of social status. For each group, a member who was intermediate in status was asked to become a confederate of the experiment and was instructed to argue either for the better site (the majority position) or for the worse site (a minority position), either early or late in the decision-making process. After the decision had been made, group members were informed that their previous ratings of each other had been lost and were asked to fill out the rating forms again. This ruse enabled the social status of the confederate to be measured before and after the decision-making event. Only confederates who argued for the *minority* position *late* in the decision-making process suffered a loss of status. I don't know how Arie and Donna managed to pull off this experiment, but it beautifully illustrates the concept of spontaneous collective intelligence. A good decision-making process begins with an alternative generating phase, followed by an evaluation phase, leading to the rejection of all but the final decision. It is appropriate to argue for minority positions early in the decision-making process but becomes inappropriate when the group is about to reach closure. The Scouts spontaneously coordinated this decision-making dance by becoming angry at the confederate only when the minority position was argued at the wrong time. Other experiments in my puzzle box showed that when group leaders are instructed to announce their own opinion early rather than late in the decision-making process, the quality of the decision suffers.

Even though "groupthink" remains a household word, I discovered that the concept had been tested and largely rejected by subsequent studies. For example, George Kennan formulated the Marshall Plan by deliberately encouraging discussion and disagreement among his advisors. By contrast, Lyndon Johnson was an overbearing leader who greeted dissenting views from his advisors with the ominous statement "I'm afraid he's losing his effectiveness," forcing many to leave his inner circle and those who remained to suppress their opinions. Historians agree that the Marshall Plan was a well-crafted decision, while Johnson's Vietnam War policy was a disaster, but what does this tell us about the quality of groups versus individuals as decision-making units? Kennan functioned as a moderator of a group process rather than imposing his own decision, while Johnson reduced the size of his decision-making group to a single person (himself). In coining the term "groupthink," Janis had made a number of general claims: that decision-making deteriorates as the group becomes more cohesive, or as the importance of the decision increases, and so on. These claims were not supported by subsequent analyses of foreign-policy decisions and laboratory experiments that manipulated group cohesion and the importance of the decision as independent variables. As one article put it, "On the basis of our review, it seems clear that there is little support for the full groupthink model.... Furthermore, the central variable of cohesiveness has not been found to play a consistent role.... This suggestion is diametrically opposed to Janis's (1982) view that high cohesiveness and an accompanying concurrence-seeking tendency that interferes with critical thinking are 'the central features of groupthink.' " These pieces of my jigsaw puzzle, comprising a dozen or so of my 495 abstracts, snapped nicely into place once I read them in detail.

Another thing that I discovered was that the comparison between individuals and groups was often nothing of the sort. Take those brainstorming experiments—please! They are based on a comparison of real groups of people bouncing ideas off each other and so-called nominal groups, whose members think in isolation. But a nominal group is not an individual; it is another kind of group that includes the investigator as one of its members. First the investigator takes the ideas compiled by the members of the nominal group and merges them into a list, throwing out redundancies along the way. The fact that this list is longer (and no less creative) than the list compiled by real groups is used as evidence that nominal groups are superior to real groups. However, the real groups do not require the services of the investigator to compile their list, and the time required by the investigator is not added to the time spent by the nominal groups. Moreover, using the list to make a decision requires examining all the items on the list. The real groups can do this while compiling the list, whereas the nominal groups can't even get started until after the experiment is over. When these factors are taken into account, it's not clear that a real group wanting to make a real decision would emulate the structure of so-called nominal groups, even for the simple tasks employed in brainstorming experiments. If we compare the performance of a single individual to either real groups or nominal groups, the result is a no-brainer. Individuals generate far, far fewer ideas. When we compiled our master list of job titles, most individuals could list only seventy or so after racking their brains, whereas together they thought of 442. The superiority of groups compared to individuals is so obvious that brainstorming researchers scarcely comment upon it in their rush to study something more subtle, such as the difference between real and nominal groups.

Sometimes even the clearest demonstrations of mental teamwork were made to appear individualistic by altering the frame of comparison. Imagine that you are a college student who studies for a test with two of your friends. After the three of you take the test individually, you are given a fourth copy to fill out as a group. Larry Michaelsen and his associates have performed this experiment in twenty-five courses taught over a five-year period, for a total of 222 groups. The average individual test score was 74.2, the average score for the best individual in each group was 82.6, and the average group score was 89.9. A total of 97 percent of the groups outperformed their best member. If this doesn't provide evidence for mental teamwork, what would? The ability of group members to evaluate and correct each other's mistakes on a case-by-case basis is even similar to the ability of our genetic machinery to detect and repair mutational mistakes in one copy of our DNA based on the "correct answer" provided by the other copy. Nevertheless, this study was criticized by other researchers who claimed that the groups would need to answer questions correctly that *every member answered wrong* to demonstrate an "assembly bonus effect." Larry and his associates replied that their groups do demonstrate an assembly bonus effect, but my point is simpler: why on earth are they setting the bar so high for something to count as a group process? Doesn't correcting each other's mistakes when somebody got it right also require a group?

By the time I finished piecing together the psychological literature, it provided compelling evidence for the advantages of thinking in groups as far as I was concerned. My analysis was published as an article with the obtuse title "Incorporating Group Selection into the Adaptationist Program: A Case Study Involving Human Decision Making." My experiments with Ralph and John were published as a second article titled "Cognitive Cooperation: When the Going Gets Tough, Think as a Group." Both are available on my Web site, and you are amply prepared to read them by now, despite the fact that they were written for a scientific audience. They dive into the details, which can be difficult or boring for some people to read, but those same details become interesting when you acquire the spirit of scientific inquiry and appreciate how they enable scientists to hold each other accountable for their factual claims.

All of the mental tasks employed by all of the psychological experiments pale in comparison to the most difficult problems of real life. A fascinating book titled *Technological Innovation as an Evolutionary Process*, edited by the distinguished physicist John Ziman, shows how many of the contrivances that make up our world are the result of group processes. If you think that the lightbulb was invented by a single person such as Thomas Edison, think again. It required dozens and dozens of people, who existed in a culture that enabled them to work as they did, which in turn was the creation of hundreds and thousands of nameless people stretching back into the mist of time. When it comes to the comparison of individuals versus groups as decision-making units, the joke is on the individual.

I Don't Know How It Works!

IN OUR HOUSE THERE are many books and few video-tapes. In fact, when our older daughter, Katie, was small, our only video was *The Wizard of Oz*, which I must have watched with her a hundred times. Somehow I didn't mind. With each viewing, I focused in more detail on those wonderful vaude-ville actors practicing their vanishing craft. Even though the movie departed from the book as only Hollywood can, it had a timelessness of its own. One of my favorite scenes came at the end, when the Wizard is departing in the balloon and Dorothy begs him to come back. "I can't!" he replies as his voice trails off in the distance. "I don't know how it works!"

That refrain deserves to be a mantra for our current under-standing of ourselves. For over a century we have tried to ex-plain as much as we can with minimalistic theories that make almost no reference to evolution. The effort to provide an evolutionary account is so new that almost everything I have related in this book was discovered within the last twenty years. These discoveries make it clear that an evolutionary ac-count is needed, but they only begin to provide the account. So much remains to be discovered in the future.

Instead of minimalistic assumptions such as the utility maximization of rational choice theory or the blank slate of behaviorism and social constructivism, we need to discover a complex psychological architecture that evolved by genetic evolution and that causes small groups to self-organize into coordinated units. Alexis de Tocqueville got it right in 1835 when he wrote, "The village or township is the only association that is so perfectly natural that ... it seems to constitute itself," but modern science has yet to even remotely take his conjecture seriously. Conscious intentional thought is just the tip of an iceberg. The rest of the iceberg operates beneath conscious awareness and must be discovered scientifically, like vision, despite the fact that it takes place within us every moment of the day. Even more strangely, it takes place *without* us, in our social intercourse in addition to our neuronal interactions. The idea that we play a role in group-level mental processes without any conscious awareness will take some getting used to, especially against the background of individualism, which has dominated the intellectual landscape for the last half century.

As if these layers of ignorance aren't enough, there is another layer that involves culture rather than genes. Our genetic architecture enables us to create, transmit, and select behaviors in roughly the same way that the immune system creates, transmits, and selects antibodies. Part of this process is conscious and intentional. To some extent, we are aware of our problems and actively seek solutions, as I just showed in the previous chapter. To a larger extent, however, the creation, retention, and selection of behaviors take place beneath conscious awareness. We learn the ways of our culture at a very young age, in the same spongelike fashion in which we learn language. As adults we adopt new behaviors and mannerisms unconsciously at least as much as consciously. Many of our current behaviors exist not because someone decided they were useful but because they outsurvived competing behaviors. Human life consists of many inadvertent social experiments. Even when we try to steer the course of events, our efforts interact with those of others in unpredictable ways that might as well have been random. A few social experiments hang together, while the others crumble into the dust. We don't even forget the failures because they weren't remembered in the first place. Like the Wizard in his ponderous hot-air balloon, we are borne skyward in cultural worlds of our own making, without knowing how they work.

To appreciate the importance of cultural evolution, let's return to the concept of our species as the primate equivalent of a social insect colony. I mentioned in Chapter 20 that insects crossed the cooperation divide only about fifteen times, but the descendants of these origination events account for over 50 percent of the biomass of all insects. This combination of rare origination followed by ecological dominance certainly describes our own species. Ed Wilson, whose name has appeared again and again in this book, for his influence on others in addition to his own contributions, made this point in a recent article written with his longtime associate Bert Holldobler and published in the *Proceedings of the National Academy of Sciences*. The article is titled "Eusociality: Origin and Consequences" and is mostly about the social insects, but it ends with the following teaser about humans: "Rarity of occurrence and unusual pre-adaptations characterized the early species of *Homo* and were followed in a similar manner during the advancements of the ants and termites by the spectacular ecological success and preemptive exclusion of competing forms by *Homo sapiens*."

This vast understatement reminds me of the way that James Watson and Francis Crick ended their famous paper on DNA: "It has not escaped our attention that the specific pairing we have postulated immediately suggests a possible copying mechanism for the genetic material." Both are single sentences that open the door to a world of implications. However, there is a difference between the ecological dominance achieved by the social insects compared to humans. After those fifteen species of insects crossed the cooperation divide, they split again and again until now there are many thousands of species of social insects. After we crossed the cooperation divide, we achieved worldwide ecological dominance while remaining a single species. The reason is that our diversification was cultural rather than genetic. There are thousands of human cultures, just as there are thousands of social insect species, but they were all created by a genetic architecture that is more or less the same around the world.

Think of it: to the best of our knowledge, we are all descended from a small population that spread out of Africa about seventy thousand years ago and proceeded to cover the globe, displacing previous populations of *Homo* and many other species as they went. This is the "ecological success and preemptive exclusion" to which Ed and Bert refer. You might celebrate or deplore it, but either way it required amazing behavioral flexibility. This was not a matter of implementing "war plans" that had previously evolved by genetic evolution in the African environment, but a more open-ended process of creating and retaining new adaptations. We diversified into thousands of "lifeways," speaking different languages and practicing different subsistence activities, from harvesting seeds to harvesting whales. We adapted to all climatic zones, from steaming jungle to parched desert to the frozen Arctic. We invaded the sea in boats. Ecologically, we became the equivalent of hundreds of different species in just a few tens of thousands of years, rather than the millions of years required for a comparable adaptive radiation by genetic evolution. Our capacity for culture shifted evolution into hyperdrive.

The advent of agriculture initiated more changes that had never before been encountered in our history as a species. Population size became limited by social organization rather than by food. Some cultures found ways to operate more efficiently and at a larger scale than others, displacing them just as earlier populations of *Homo* had been displaced. Sometimes the displacement took the form of violent conquest, but it also took the form of imitation and assimilation. With cultural evolution, the battles do not necessarily take place on actual battlefields. Ever-increasing scales of social organization carried us into the period that we myopically call human history, where it continues to this day, as world historians such as William McNeill appreciate.

If we take our own balloon trip over the Ivory Archipelago, we can find some outstanding examples of cultural evolution in action. Let's touch down in the upper Nile basin of Africa during the early 1800s, inhabited by a tribe called the Dinka except for a small region occupied by another tribe called the

Nuer. Both tribes raised cattle and grew millet in the same kind of physical environment, yet the Nuer vastly expanded their territory at the expense of the Dinka and might well have completely replaced them except for the appearance of a disease that devastated the cattle stocks of both tribes in the late 1800s, followed by the intervention of the Anglo-Egyptian colonial administration in the early 1900s. The Nuer expansion has been studied by anthropologists for over sixty years, making it one of the best-documented examples of how cultures replace each other. Raymond Kelly's *The Nuer Conquest* provides a good single source for this literature, which shows how thorough and scientifically grounded the study of cultural evolution can be.

Tribes are not homogenous units; they consist of subtribes, which in turn consist of clans, with variation among units at all scales. The Nuer were a subtribe of the Dinka that became distinctive enough to become a tribe in its own right. Their language and most of their customs were similar, with some key differences that *made* a difference as far as their interactions were concerned. Marriage required a payment of cattle to the bride's family for both tribes, but the payment was larger and required more male cattle for the Nuer than the Dinka. This seemingly trivial difference *made* a difference because it caused the Nuer to manage their herds differently, overgrazing their land and providing an incentive to invade Dinka territory for cattle, land, and women.

The two tribes also differed in their kinship systems. In the Dinka, people who considered themselves kin were concentrated in settlements where the business of raising cattle and growing millet took place. The Nuer kinship system linked people from different settlements that had no reason to interact—except when it came time for a cattle raid. The Nuer were consistently able to field a larger fighting force than the Dinka by virtue of their kinship system. Social obligations among kin gave the Nuer an advantage in between-group competition, regardless of whether the "kin" were actually genetically related to each other.

With a greater incentive and larger fighting capacity, the Nuer simply nibbled away at Dinka territory in dozens of separate raids until their territory had increased fourfold by 1880. There were deaths, of course, but many more Dinka were appropriated into Nuer society as wives and slaves. Even slaves rapidly changed their social identity. When the great British anthropologist E. E. Evans-Pritchard was studying the Nuer in the 1940s, at least half of the Nuer population was formerly Dinka, and it was difficult to distinguish a second-generation Dinka from a more "blueblood" Nuer. One *culture* was replacing another, but this was not just a matter of *individuals* replacing each other.

I wish that I could dwell more on this fascinating example, but even my brief account reveals some important facts about cultural evolution. The first ingredient of an evolutionary process is *variation*, as I showed in Chapter 3. The Nuer example shows that cultural variation is seldom in short supply. It seems to emerge spontaneously from human social interactions, creating offshoots such as the Nuer from a previously existing society such as the Dinka. This kind of spontaneous generation might seem remarkable but it is actually expected for all complex systems, including purely physical systems such as the weather. The weather is famously unpredictable because tiny changes in initial conditions become amplified over time by complex interactions. This is often called the butterfly effect, in humorous reference to the possibility that a butterfly flapping its wings over Brazil can set off a tornado in Texas. Given the unpredictability of the weather, it shouldn't surprise us that small cultural differences among social groups can become amplified into larger differences over time.

The second ingredient of an evolutionary process involves the *consequences of variation*. Differences in how people behave must *make* a difference in terms of survival and reproduction. The Nuer example shows how such things as marriage customs and kinship systems can have a huge impact on competitive interactions among cultures. At the same time, it would be a mistake to think that every difference between the Nuer and Dinka has functional significance. There is more to evolution than natural selection, as I was careful to stress in Chapters 6 to 8, and those lessons apply as forcefully to cultural evolution as to genetic evolution. Raymond Kelly discusses numerous features of Nuer and Dinka society that are the cultural equivalent of mad monkeys and dogs' curly tails.

The third ingredient of an evolutionary processes is *heritability*. Behaviors must perpetuate themselves through time for differences in survival and reproduction to have a cumulative effect. The third ingredient might seem to conflict with the first ingredient. If tiny differences can change the course of history for a culture, where is the stability required to preserve successful solutions? The Nuer example shows that cultures possess stability in addition to variability, as paradoxical as that might sound. The Nuer culture was an offshoot of the Dinka culture, but it was so stable that it could withstand a massive influx of Dinka individuals without changing! Another example of cultural stability revealed itself when the cattle herds were reduced by disease in the late 1800s. Most Nuer no longer had the number of cattle prescribed by their culture to purchase a new wife. Marriages still took place, but the *ideal* remained unchanged and reasserted itself as soon as the stocks recovered. As Kelly put it: "Although fluctuations in the size of Nuer herds engender variations in the magnitude of bridewealth payment ... such fluctuations do not produce a redefinition of the acceptable and ideal payments that define it. These are very deeply embedded in kinship obligations." The culture provides a stable set of guidelines for how to behave, even when people cannot live up to the guidelines in reality.

In short, the Nuer example shows how the fundamental ingredients of an evolutionary process can exist for cultural in addition to genetic change, but it is most remarkable for showing how cultural evolution takes place beneath conscious awareness. Both the Nuer and the Dinka doubtless thought very hard about matters of importance in their everyday lives, such as raising cattle, growing millet, getting married, and planning raids, but they appeared completely unaware of the larger cultural parameters that framed their everyday decisions. There is not a shred of evidence, for example, that the Dinka ever wondered about why they kept losing ground to the Nuer or how they could change their customs to become more competitive. Their ignorance of their own culture was as profound as their ignorance of their genetically evolved adaptations.

Let us now leave the Nuer and travel to the time in human history when writing emerged as a cultural adaptation. Our guide will be Walter J. Ong, a Catholic priest and world-famous scholar of the humanities who died at the age of ninety in 2003. In his book *Orality and Literacy*, Ong persuasively argued that writing fundamentally changed the character of human thought. In purely oral cultures, everything known must be remembered by repeating it again and again, placing enormous constraints on the organization of knowledge and communication. Writing enabled us to store information outside of our heads as never before (art and physical artifacts might perform similar functions), freeing our minds for new activities. According to Ong, it is no accident that the flowering of philosophy, logic, mathematics, and other forms of analytic thought in ancient Greece took place when it did. These mental operations were luxuries that no one could afford prior to the advent of writing.

Today, human thought has been so transformed that members of literate societies can scarcely imagine what it was like to think without the help of the written word. Ong puts it this way: "You know what you can recall. When we say we know Euclidean geometry, we mean not that we have in mind at the moment every one of its propositions and proofs but rather that you can bring them to mind readily. We can recall them. The theorem 'You know what you can recall' applies also to an oral culture. But how do persons in an oral culture recall? The organized knowledge that literates today study so that they 'know' it, that is, can recall it, has, with very few if any exceptions, been assembled and made available to them in writing. This is the case not only with Euclidean geometry but also with American Revolutionary history, or even baseball averages and traffic regulations."

Knowledge in oral societies is stored and communicated largely in the form of proverbs. We still make use of proverbs as memorable nuggets of information that are especially apt for a given situation, such as "A stitch in time saves nine," but they play a much smaller role in our thoughts and conversation than in oral cultures. Ong continues: "Formulas help implement rhythmic discourse and also act as mnemonic aids in their own right, as set expressions circulating through the mouths and ears of all. 'Red in the morning, the sailor's warning; red in the night, the sailor's delight.' 'Divide and conquer.' 'To err is human, to forgive is divine.' 'Sorrow is better than laughter, because when the face is sad the heart grows wiser' (Ecclesiastes 7:3). 'The clinging vine.' 'The sturdy oak.' 'Chase off nature and she returns at a gallop.' Fixed, often rhythmically balanced, expressions of this sort and of other sorts can be found occasionally in print, indeed can be 'looked up' in books of sayings, but in oral cultures they are not occasional. They are incessant. They form the substance of thought itself. Thought in any extended form is impossible without them, for it consists in them."

The difference between oral and literate thought is wonderfully captured in the novel *No Longer at Ease*, by Chinua Achebe, which is set in Nigeria in the 1960s. The protagonist, Obi Okonkwo, is a young man who has been sent to England by his village to receive a modern education, so that he can return and obtain a high post in the Nigerian government. He has made the transition to literate thought and proudly regards himself as a Nigerian instead of a son of Umuofia, his village, but the members of the village who sacrificed so heavily to give him his education still regard him as one of their own. The conflict and ultimately tragic ending form the basis of the book, but the dialogue of the villagers also illustrates how members of oral societies think and talk to each other in terms of proverbs. In the following passage, Obi (who is the grandson of Ogbuefi Okonkwo, the protagonist of Achebe's most famous novel, *Things Fall Apart*) is visiting his village for the first time since returning from England:

Four years in England had filled Obi with a longing to be back in Umuofia. This feeling was sometimes so strong that he found himself feeling ashamed of studying English for his degree. He spoke Ibo whenever he had the least opportunity of doing so. Nothing gave him greater pleasure than to find another Ibo-speaking student in a London bus. But when he had to speak in English with a Nigerian student from another tribe he lowered his voice. It was humiliating to have to speak to one's countryman in a foreign language, especially in the presence of the proud owners of that language. They would naturally assume that one had no language of one's own. He wished they were here today to see. Let them come to Umuofia now and listen to the talk of men who made a great art of conversation. Let them come and see men and women and children who knew how to live, whose joy of life had not yet been killed by those who claimed to teach other nations how to live.

This "great art of conversation" consisted of the artful stringing together of proverbs, which themselves are as artful as polished stones: "Her stay with him is like the stay of the moon in the sky. When the time comes she will go." "We have our faults, but we are not empty men who become white when they see white, and black when they see black." "The great tree chooses where to grow and we find it there, so it is with greatness in men."

The gulf between literate and oral thought is also illustrated by a series of interviews that the great Russian psychologist Alexander Luria conducted with illiterate peasants in the 1930s. The interviews were conducted in the unthreatening atmosphere of a tearoom. While sipping tea across a table might have helped to decrease social distance, it could not diminish the mental distance between Luria as a member of a literate society and his guests as members of a peasant society that remained primarily oral. When asked to identify geometric figures, the peasants never responded with abstract terms such as "circle" or "square" but with practical names such as "plate" or "mirror." When Luria presented them with drawings of a hammer, saw, hatchet, and log, asking them which three belonged to one category and the fourth to another, the peasants were unable to separate the three tools from the log. Instead they kept thinking about using the tools on the log: "They're all alike. The saw will saw the log and the hatchet will chop it into small pieces. If one of these has to go, I'd throw out the hatchet. It doesn't do as good a job as a saw."

The peasants seemed especially mystified by questions about themselves, as the following transcript reveals:

> Question: "What sort of person are you? What's your character like? What are your good qualities and short-comings? How would you describe yourself?"
>
> Answer: "I came here from Uch-Kurgan, I was very poor, and now I'm married and have children."
>
> Question: "Are you satisfied with yourself or would you like to be different?"
>
> Answer: "It would be good if I had a little more land and could sow some wheat."
>
> Question: "What are your shortcomings?"
>
> Answer: "This year I sowed one pood of wheat, and we're gradually fixing the shortcomings."
>
> Question: "Well, people are different—calm, hot-tempered, or sometimes their memory is poor. What do you think of yourself"
>
> Answer: "We behave well—if we were bad people, no one would respect us."

Another peasant eloquently responded to the last question by replying, "What can I say about my own heart? How can I talk about my character? Ask others; they can tell you about me. I myself can't say anything."

These responses reflect the overwhelming need of people in oral societies to use their heads for practical purposes. Abstractions such as "circle," "tool," or "the type of person I am" are luxuries that become useful only in the context of a society organized by writing. As Ong puts it: "Without a writing system, breaking up thought—that is, analysis—is a high-risk procedure."

Members of oral societies are as intelligent as anyone else, but they appear retarded by literate standards. In the follow-ing passage, Ong nicely shows how a literate person can also

appear retarded by oral standards: "Proponents of intelligence tests need to recognize that our ordinary intelligence
test questions are tailored to a special kind of consciousness,
one deeply conditioned by literacy and print. . . . A highly intelligent person from an oral or residually oral culture might
be expected normally to react to Luria's type of question, as
many of his respondents clearly did, not by answering the
seemingly mindless question itself but by trying to assess the
total puzzling context (the oral mind totalizes): What is
he asking me this stupid question for? What is he trying to
do? 'What is a tree?' Does he really expect me to respond to
that when he and everyone else has seen thousands of trees?
Riddles I can work with. But this is no riddle. Is it a game? Of
course it is a game, but the oral person is unfamiliar with the
rules. The people who ask such questions have been living in
a barrage of such questions from infancy and are not aware
that they are using special rules."

Just like the Nuer and the Dinka, literate societies are a
product of cultural evolution in addition to genetic evolution, and their members are profoundly ignorant about both.

The next and final stop on our tour is the United States of
America. Our guide is Richard Nisbett, an eminent social
psychologist whose interest in cultural evolution was sparked
by a youthful experience, much like Bill McNeill's interest in
dance described in Chapter 24. Born and raised in Texas,
Dick Nisbett moved to Massachusetts at the age of eighteen
and couldn't help but notice the cultural difference. The
South is known for its gentility and hospitality, but also for its
violence. Feuds, duels, and lynchings have been primarily
southern affairs throughout American history. The famous
feud between the Hatfields and McCoys that took place on
the Kentucky/West Virginia border would seem out of place
in New England. According to one estimate, the homicide
rate for the plateau region of the Cumberland Mountains of
Tennessee between 1865 and 1915 was more than ten times
today's national average and twice as high as our worst inner cities. Southerners take a much greater interest in war
and have been overrepresented in the military throughout
American history. Even entertainment is more violent in the

South. If you visit a southern bar, you might witness a game called "purring" in which two men grasp each other by the shoulders and start kicking each other until one gives up.

After establishing his reputation as a general social psychologist with books such as *Human Inference: Strategies and Shortcomings of Social Judgment* (co-authored in 1985 with Lee Ross), Dick decided to focus professionally on cultural variation, starting with the North-South difference that intrigued him as a youth. He and his graduate student Dov Cohen pieced together multiple strands of evidence that are summarized in their 1996 book *Culture of Honor: The Psychology of Violence in the South*, which for me is as entertaining as any mystery novel.

Southern violence has been variously attributed to poverty, climate, and the institution of slavery, which sanctioned violence as a form of social control in the past. These factors might play a role, but they ignore a fourth factor that has nothing to do with American history. The people who colonized the North and South *were already different* in their cultural practices before they set foot on American soil. The North was colonized primarily by farmers and merchants who were accustomed to having their affairs governed by strong religious and political authority, such as the Puritans, who colonized New England, and the Quakers, who colonized Pennsylvania and Delaware. The South was colonized in part by the sons of British noble families who established plantations in the fertile lowlands, but even more by herding people from the highland regions of the British Isles, such as the Scotch-Irish, who moved into the upland regions of the American South and continued living much as they had before.

Herding people all around the world share a common set of problems. Unlike a farmer's field, their property is mobile and can be easily stolen. They also tend to live in sparsely populated areas that are difficult to govern by centralized authority. The only solution to these problems is self-defense, leading to a "culture of honor" that anthropologists have documented for people as historically distinct as the Nuer and Dinka in Africa, Greek shepherds, the Navaho of the

American Southwest, and the European Celts, whom the Romans respected for their ferocity but disparaged for their lack of organization. These people are similar to each other not because they are historically connected but because cultural evolution has caused them to converge on a common solution to a common set of problems.

In a culture of honor, using violence to defend one's reputation is not just morally acceptable but imperative. The journalist Hodding Carter recalls serving on a jury in Louisiana in the 1930s. The case involved a man who lived next to a gas station, where the hangers-on had been teasing him. One day he opened fire on them with a shotgun, wounding two and killing an innocent bystander. Carter was the only member of the jury who proposed a verdict of guilty. The others protested: "He ain't guilty. *He wouldn't have been much of a man if he hadn't shot them fellows.*"

For children, the culture of honor is the only world that they know. Chris Boehm, whose work on egalitarianism I described in Chapter 21, spent his early career studying the herding people of Montenegro, which is very much a culture of honor. He once told me about an evening family gathering in which a little boy, barely able to walk, was given a fireplace poker and taunted until he attacked his adult assailants in a rage, to the merriment and encouragement of all. Similarly, an observer of southern life commented that small children "were supposed to grab for things, fight on the carpet to entertain parents, clatter their toys about, defy parental commands, and even set upon likely visitors in a friendly roughhouse." As boys grew older, these games became training grounds for combat. A boy who came home complaining about a bully would be sent back to show the bully "what you are made of." A boy who dodged a stone would be called a coward; the idea was to allow oneself to be hit and then respond in kind.

Women play an integral role in the culture of honor, sometimes in the actual fighting but especially in their influence on the men. Here is how a Roman described the women who were the ancestors of the Scotch-Irish: "A whole troupe of foreigners would not be able to withstand a single Gaul if he called his wife to his assistance, who is usually very strong and with blue eyes; especially when, swelling her neck, gnashing her teeth, and brandishing her sallow arms of enormous size, she begins to strike blows mingled with kicks, as if they were so many missiles sent from the string of a catapult."

The southerners of early America idolized the mothers of ancient Sparta, who were said to command their sons to return from battle either with their shields or upon them. Sam Houston's mother gave him his musket with the words "Never disgrace it; for remember, I had rather all my sons should fill one honorable grave, than that one of them should turn his back to save his life." She then gave him a plain gold ring with the word "honor" engraved inside it. When a southern veteran of the Civil War was asked why the Confederates kept on fighting after defeat was certain, he replied, "We were afraid to stop.... Afraid of the women at home ... They would have been ashamed of us."

The modern South is no longer a herding society. The vast majority of people have workaday jobs, just as in the North, but that doesn't mean that the cultural differences between the North and South are going to disappear anytime soon. Cultural evolution is faster than genetic evolution but still requires time to respond to environmental change. Remember that conscious awareness and intentional planning are just the tip of an iceberg. The values and practices that are learned and transmitted beneath conscious awareness have much more inertia. We dance with the ghosts of past environments for cultural evolution, no less than genetic evolution.

A set of experiments performed by Dick, a transplanted Texan, and Dov, a native northerner, shows just how much culture can endow a person with a second nature. The experiments were conducted on male undergraduate students at the University of Michigan and had the quality of a *Candid Camera* stunt. One by one, each student was asked to fill out a questionnaire and take it to a table at the end of a long narrow hallway. On his way, he passed a filing cabinet and a confederate of the experiment, who had to stop what he was doing and close the file drawer to allow the subject to pass. After dropping off the questionnaire, the subject returned

down the same hallway and again interrupted the confederate, who stood up, slammed shut the file drawer, and marched
past the subject, bumping him on the shoulder and calling
him an "asshole."

The purpose of this carefully staged insult was to see if
University of Michigan students who come from southern
states respond differently than students who come from
northern states. Students from both regions came from affluent families, with a median income of $85,000 for the
northerners and $95,000 for the southerners. Southern stu-
dents who decide to attend a northern university are probably more like northerners than those who stay at home.
Nevertheless, there were huge differences in how southern
and northern students responded to the insult. Southerners
tended to become visibly angry, while northerners tended to
respond with amusement, according to observers who were
stationed on each end of the hall and did not know the identity of any particular subject. In an even more elaborately
stage-managed version of the experiment, the insult was followed by the appearance of a second confederate walking
down the hall who happened to be six feet three inches
tall and weighed 250 pounds. How close would the insulted
subject allow the second confederate to approach before
stepping aside? Northerners stepped aside at a distance of
about five feet, regardless of whether they had been insulted.
Southerners stepped aside at the very respectful distance of
nine feet in the absence of an insult (that famous southern
politeness) but belligerently waited until a distance of three
feet following an insult. Perhaps the most telling version of
the experiment involved obtaining a sample of saliva before
and after the insult for hormonal analysis. Levels of cortisol
and testosterone (two hormones associated with stress and
aggression) soared for southerners but not northerners after
the insult. There were no differences before the insult or in
the no-insult control condition. Even affluent southern students who had elected to attend a northern university were
behaviorally and hormonally primed by their culture to defend their honor.

Dick can relate to the southern students because he remains

programmed by the southern culture of his youth, even though he has been living in the North for several decades. He once told me that if he sees someone trespass on his lawn from his upstairs window, before he knows it he is outside defending his turf. It doesn't matter that he's middle-aged, out of shape, and in his pajamas. "It's in my bones!" he said. But it's *not* in his bones, if by that we mean in his genes. The *capacity* for that response is in his genes, along with my own incapacity as the product of the Puritan culture of New England, and a vast range of other capacities, past and future. To understand *which* capacity is "in the bones" of any particular individual or society, we must look to the process of cultural evolution, not genetic evolution.

Our tour has included three outstanding examples of cultural evolution, but you might want to extend the tour on your own. Read Dick Nisbett's most recent book, *Geography of Thought*, to learn about differences between Asian and Western societies, Robert Putnam's book *Making Democracy Work* to learn about cultural variation within the nation of Italy, or Francis Fukuyama's book *Trust* to learn about international cultural variation in the capacity for trust. Above all, read Alexis de Tocqueville's classic *Democracy in America*, if you haven't already, for an amazingly penetrating analysis of the causes and consequences of cultural variation.

It might seem odd that I began this chapter by stressing our ignorance about cultural evolution, but then proceeded to describe well-known and well-documented examples. Raymond Kelly, Walter Ong, Dick Nisbett, Robert Putnam, Francis Fukuyama, and Alexis de Tocqueville are all giants within their respective fields and do not require an introduction by an itinerant evolutionist such as myself. Or do they? It all depends upon which island of the Ivory Archipelago we visit. A giant in one discipline can be unheard-of in another. Even worse, the *ideas* that a giant represents can be foundational for one discipline, totally absent for another, and heretical for still another. Academic disciplines are cultures too, with certain propositions "in their bones" and others beyond their imagination. Earlier in his career, Dick Nisbett and most of his colleagues assumed that whatever they discovered for American college students would represent the psychology of our entire species. He was shocked to discover the importance of culture and has become a powerful but minority voice within his own field of social psychology.

Even my own evolutionist colleagues who are well versed in genetic evolution often don't know what to make of cultural evolution. Give them a piece of modern human behavior or psychology to explain, and they try to imagine how it would have functioned in the Stone Age. This is a reasonable way of thinking about some human adaptations, as I have stressed in earlier chapters of this book, but it completely ignores the more open-ended process of cultural evolution. At present there is no consensus among evolutionists about what counts as part of our genetically innate psychological architecture and what counts as a culturally evolved product of the architecture.

When we travel to islands of the Ivory Archipelago that are most appreciative of cultural variation, we find that they are also most resistant to the idea of evolution in any form, as I described in Chapters 1 and 24. Many would deny the existence of a complex genetic architecture that is culturally universal, or the possibility of explaining cultural diversity in the same way that evolutionists explain biological diversity. If there is a consensus about evolution among connoisseurs of cultural variation, it is an erroneous one.

These are not just matters of idle intellectual curiosity. Our future depends upon adapting our cultures to the realities of modern life at an unprecedented spatial and temporal scale. The idea that we can do this without a detailed knowledge of genetic and cultural evolution will appear laughable in retrospect—if we are lucky enough to persist in our current maladaptive cultures for so long. Perhaps someday we will confidently steer ourselves into the future like Captain Kirk and the starship *Enterprise*. Until then, we will be like the Wizard of Oz in his hot-air balloon, who doesn't know how it works.

28

Darwin's Cathedral

IT WAS JANUARY 2002 and the first day of a new course that I was teaching on evolution and religion. Thirty-six students had enrolled. As they introduced themselves to me and each other, I learned that nineteen were Christian, nine were Jewish, three were Muslim, one was Hindu, one was Buddhist, and three were nonbelievers. Isn't it amazing, I thought, that in a class on all of the religions that have existed around the world and throughout history, the students in this very room came from all around the world and represented most of the religious traditions that we would be studying.

The main text for the course was my newly written book *Darwin's Cathedral: Evolution, Religion, and the Nature of Society. Darwin's Cathedral* attempts to do for religion what I and my evolutionist colleagues have done for so many other subjects described in this book, from infanticide to cultures of honor. Many different religions are discussed, but most of them were chosen selectively, leaving me open to the criticism of bias. Just as statistics can be selectively reported to support almost any position, perhaps I merely sifted among many religions to find those that supported my own pet ideas.

There is a solution to the problem of selection bias, and it is called random sampling. The idea is simplicity itself: study a sample of religions chosen without respect to any particular hypothesis and let the chips fall where they may. In *Darwin's Cathedral* I initiated this procedure using the sixteen-volume *Encyclopedia of World Religions*, edited by the great religious scholar Mircea Eliade. I wrote a computer program that chose a volume number at random, followed by a page number at random within the volume. I then evaluated the encyclopedia entry that included the page number to see if it referred to a single religion, defined as "a recognizable group of people with beliefs and practices that can be distinguished from other beliefs and practices." This definition distinguishes among denominations within a major religious tradition, such as Calvinism and Lutheranism, in addition to different major traditions, such as Christianity and Buddhism. If an entry satisfied the definition, then the religion was included in my sample. If not (some entries referred to general topics such as "myth" or "polytheism" rather than a single religion), then I paged forward to the first appropriate entry. This procedure was repeated to give me a sample of religions, chosen without respect to any pet idea of mine. All beliefs and practices were assumed to be religious because, after all, they were taken from an encyclopedia of religion. In other words, I let Eliade and his team of editors decide what counts as religious, rather than imposing my own definition.

In *Darwin's Cathedral*, I chose religions by this method to illustrate the principle of random sampling. The purpose of my course was to actually study the religions in the sample with the help of the students. After they introduced themselves, I explained that each student would be assigned a single religion from the sample to research over the course of the semester. The assignment would be random, of course, so that a Jewish student might be assigned a Muslim religion, a Christian student might be assigned a Hindu religion, and so on. Their challenge was to become experts on their particular religion by scouring the scholarly literature. Then they would write an extended essay on their religion consisting of answers to thirty-two questions that had been prepared by me. I would read the essays partway through the course and send the students back to the literature to gather more information if necessary. Finally, we would use methods developed by social scientists to convert these mountains of descriptive information into numerical form.

The class fell silent as I outlined the objectives of the course, as if they thought they were joining the Boy Scouts and now found themselves facing a drill sergeant in a military boot camp. Nevertheless, they were interested and nobody elected to drop the class, as they easily could at the beginning of the semester. After all, it's exciting to think that you can actively contribute to knowledge rather than passively receive it, especially for a topic as fascinating and important as religion.

My own fascination with religion began three years earlier when I encountered the following passage, written over 350 years ago by a member of the Hutterite faith (a Christian Anabaptist sect):

> *True love means growth for the whole organism, whose members are all interdependent and serve each other. That is the outward form of the inner working of the Spirit, the organism of the Body governed by Christ. We see the same thing among the bees, who all work with equal zeal gathering honey.*

You might have wondered at the very beginning of this book about my tall claim that "evolution and religion, those old enemies who currently occupy opposite corners of human thought, can be brought harmoniously together." Now perhaps you can see how a union is possible. The last eleven chapters have been all about groups as organisms, from the origin of life to multicellular bodies, social insect colonies, and our ancestors, who became so different from other primates by crossing the cooperation divide. The writer of this passage knew nothing about evolution or science, but his comparison of bodies, beehives, and his own religious group struck me as much more than a poetic metaphor. If I really wanted to study human groups as comparable to bodies and beehives, shouldn't I be studying religious groups? After all, that is how at least some religious believers describe themselves!

I now had the interest, but could I really afford to commit the time and effort required for such an ambitious project? A scientist such as myself is expected to fund his research with grants. I knew that federal granting agencies such as the National Science Foundation would be far too timid to fund a proposal on religion from an evolutionary perspective. Fortunately, a private foundation created by the wealthy investor John Marks Templeton was dedicated to the study of religion from a scientific perspective. Sir John (he was knighted by Queen Elizabeth II in 1987) was born into a Presbyterian family in rural Tennessee, not far from where the famous Scopes "Monkey Trial" took place when he was twelve. Still active at the age of ninety-two, he has been guided by religion throughout his life, unabashedly starting his business meetings with a prayer. Unlike the creationist neighbors of his youth, however, Sir John thinks that science and religion have much to offer each other. In particular, he thinks that religious traditions around the world are storehouses of wisdom that can be further enlightened by scientific research. The mission of his foundation is to provide the funds to make such scientific research possible. At the same time that I was becoming interested in religion from an evolutionary perspective, the Templeton Foundation announced a new initiative for funding scientific research on forgiveness. I teamed up with my friend and colleague Chris Boehm, whose work on egalitarianism was described in Chapter 21, and we wrote a grant to study forgiveness in hunter-gatherer societies (Chris's part) and modern religions (my part) from an evolutionary perspective. The project was funded, and I was launched on one of the most exciting intellectual adventures of my career.

My own background is not at all religious. My parents were both warm, nurturing, and moral people, but neither attended church, and my father in particular was scornful of religion. He was once visited by a local minister who said, "Won't you join our congregation, Mr. Wilson? Our average member earns a salary of over $20,000," a lot of money in those days. I can still remember my father's gleeful expression as he told me that story, as if the hypocrisy of all religion had been revealed.

Armed with little by way of personal experience, I set out on a tour of the Ivory Archipelago to see what I could learn about religion. This was a bit like Bilbo Baggins of J. R. R. Tolkien's *The Hobbit* leaving the Shire to discover that the world is much larger than he imagined. The literature on religion is vast. First there are the religious texts, which stretch back thousands of years. Then there is the literature that scholars of all sorts have written about religion. To give you an idea of its vastness, my university is not strong in the area of religion and our library is modest by any standard, yet it contains over a hundred books on the particular version of Christianity founded by John Calvin in the 1500s. The library at Calvin College in Grand Rapids, Michigan, a stronghold of Calvinism in America, has well over ten times that number. Every notable version of Christianity has received similar attention, not to speak of the other major religious traditions. Finally, there is a more recent social scientific literature on religion, in scores of books and journals such as the *Journal for the Scientific Study of Religion*. It is absurd to think that I or anyone else could master such a large body of information, in three years or a lifetime.

My challenge, however, was not to assimilate all of this information but to provide a theoretical framework for organizing it. What *should* religions look like from an evolutionary perspective? My main hypothesis was that religious groups are products of cultural group selection and are indeed like bodies and beehives. A given religion adapts its members to their local environment, enabling them to achieve by collective action what they cannot achieve alone or even together in the absence of religion. The primary benefits of religion take place in this world, not the next. Even though elements of religion often appear bizarre, irrational, and downright dysfunctional to nonbelievers, when examined closely most of them will make sense as part of a "social physiology" that coordinates action and solves the all-important problem of cheating from within.

On the other hand, perhaps religion is primarily a product of within-group selection rather than between-group selection. When people are encouraged to obey the Golden Rule and sacrifice in this world to receive their reward in heaven, perhaps they are being deceived by their own leaders, consciously or unconsciously. If so, then when we look closely at religion it should stand exposed as a scam operation, with the leaders fleecing rather than leading their flocks. There are certainly many examples of abuse and exploitation in the history of religions, such as the practices of the Catholic Church that led to the Protestant Reformation and the Catholic Counter-Reformation, but perhaps Karl Marx was right when he said that *most* elements of religion are "the opium of the masses."

The fact that cultural evolution involves passing ideas from person to person raises a third possibility. Disease organisms such as herpes and the cold virus also pass from person to person. They benefit not individuals or groups but only themselves. My evolutionist colleague Richard Dawkins has famously suggested that bits of culture called "memes" can evolve to be parasitic organisms in their own right. If so, then a close look at religious movements might reveal them to be like disease epidemics that leave everyone worse off than before, leaders and followers alike.

Or perhaps our enchantment with religion is a form of dancing with ghosts, as I described in Chapter 8. Just as our eating habits made sense in ancient but not modern environments, perhaps our impulse to help others made sense in ancient microsocieties composed of genetic relatives but not the mega-societies of today. If so, then a close look at religion would reveal it to be like obesity, something that we do because we can't help it, even though it is no longer good for us.

Or perhaps religions are like mad monkeys and a dog's curly tail, which have no function and persist only by virtue of a connection to something else that does. After all, many traits associated with religion are also expressed in non-religious contexts. Perhaps their nonreligious expression is clearly adaptive and their religious expression reflects a cost that can't be avoided. If so, then we must look outside of religion to explain the existence of religion as a by-product.

In short, evolutionary theory provides not one but at least five major hypotheses about religion. All of them are plausible and might potentially explain some or all features of religion. The few evolutionists who have thought about religion do not agree about which hypotheses are more likely to be correct; I favor the group-level adaptation hypothesis, but Richard Dawkins and philosopher Daniel Dennett favor the cultural parasite hypothesis in their books *The God Delusion* and *Breaking the Spell.* Scientists aren't expected to agree *prior* to testing their favored hypotheses. The agreement comes after some hypotheses explain the facts of the real world better than others. At this stage of my inquiry, I was merely trying to articulate the major evolutionary hypotheses about religion in preparation for testing them.

Most theories of religion were formulated by people who never used the E-word, but I quickly discovered that they could be categorized in terms of the five major evolutionary hypotheses without requiring additional hypotheses. For example, the great sociologist Émile Durkheim (who was the son of a rabbi) defined religion this way in 1912: "A religion is a unified system of beliefs and practices relative to sacred things ... which unite into one single moral community called a Church, all those who adhere to them." Durkheim definitely regarded religious groups as like bodies and bee-hives (the first major hypothesis outlined above), but he provided no evolutionary explanation, and his successors classified his theory as "sociological" rather than "evolution-ary," a term that they associated with another set of ideas about religion. These configurations of ideas must have seemed reasonable at the time but they no longer made sense against the background of modern evolutionary theory. Even worse, I discovered that Durkheim's views on religion had been largely rejected by modern social scientists along with the larger tradition of functionalism (the concept of society as like an organism) that he represented. I loved reading Durkheim's *Elementary Forms of Religious Life* and at one point became so excited that I exclaimed to my wife: "Anne! Durkheim is speaking to me across the chasm of time!" Anne is used to such utterances and smiled indulgently before re-turning to her work. I thought that Durkheim was on the right track and could be given a modern evolutionary formulation, but I could see that I would have a tough time convincing some of the experts.

Current social scientists rely primarily upon economic theory rather than Durkheim to explain the nature of religion. It didn't take me long to discover the work of Rodney Stark and William Bainbridge, whose many books include *A Theory of Religion* and *The Future of Religion*. According to their theory, people are very good at forming explanations about the world around them and reasoning in terms of costs and benefits to get what they want. Unfortunately, some things simply cannot be had, such as rain during a drought or everlasting life. Their unavailability doesn't prevent us from wanting them, however, so we invent religions in a futile effort to get what we can't have. We pray to God for everlasting life, not to convey us to work in the morning. Religion per se does not produce any practical benefits. It is a *by-product* of the economic mind, which produces obvious benefits in nonreligious contexts but merely spins its wheels in the case of religion. According to Stark: "All aspects of religion—belief, emotion, ritual, prayer, sacrifice, mysticism, and miracle—can be understood on the basis of exchange relations between humans and supernatural beings."

For Stark and Bainbridge, the *E*-word is "economics" rather than "evolution," but these should not be regarded as alternative theories. Most economists cheerfully admit that the human ability to form explanations and reason on the basis of costs and benefits evolved by genetic evolution. They regard economic theory as consistent with evolutionary theory in a way that doesn't require much knowledge about evolutionary theory, enabling them to proceed on the basis of their minimalistic assumptions. I have criticized this position in previous chapters, but even so it is easy to classify Stark and Bainbridge's theory as a by-product explanation of religion (the fifth major hypothesis outlined above). Two of my evolutionist colleagues, Pascal Boyer in his book *Religion Explained* and Scott Atran in his book *In Gods We Trust*, differ from Stark and Bainbridge in their conception of the human mind but also regard religion as a by-product of psychological adaptations that evolved in nonreligious contexts. If they are on the right track, then the fifth major hypothesis will emerge triumphant and my own favored first major hypothesis will go down in flames when the facts about religion are consulted.

The first two chapters of *Darwin's Cathedral* are devoted to this kind of conceptual groundwork, but the real fun and suspense of writing the book began with my attempt to describe actual religions in detail. How well would my evolutionary framework make sense of the facts and which (if any) of the major hypotheses would be supported? I decided to begin with Calvinism, one of many versions of Christianity, because it represented a natural experiment. The time was the 1530s in the chaotic days of the Protestant Reformation. The city of Geneva had recently expelled the Roman Catholic Church as part of becoming independent from the duchy of Savoy. It craved independence but was totally dependent upon the Swiss Confederacy for military support, especially the city of Berne. The Swiss Protestant reform movement had spread to Geneva but lacked organization. In addition, the city was governed by a democratically elected council that had only recently gained independence from the Catholic Church and was not about to yield to a new religious authority.

Into this volatile political and religious environment came John Calvin, a bookish scholar who was merely passing through but who was shamed into staying by Geneva's two leading reformers. It is worth quoting Calvin's own account of the historic event:

A little while previously, popery had been driven out by the good man I have mentioned [Guillaume Farel] and by Pierre Viret. Things, however, were still far from settled, and there were divisions and serious and dangerous factions among the inhabitants of the town.... Farel (who burned with a marvelous zeal to advance the gospel) went out of his way to keep me. And after having heard that I had several private studies for which I wished to keep myself free, and finding that he got

nowhere with his requests, he gave vent to an impreca-
tion, that it might please God to curse my leisure and
the peace for study that I was looking for, if I went away
and refused to give them support and help in a situation
of such great need. These words so shocked and moved
me, that I gave up the journey I had intended to make.

By Calvin's own account, Geneva was unable to hang together as a social unit despite a strong democratically elected
government. Historians and Calvin's contemporaries agree:
factionalism was evidently called "the Genevan disease."
Calvin and Farel outlined a religious agenda that included a
catechism and a short set of rules called the Ecclesiastical
Ordinances. At first the city fathers were so alarmed by its
austerity that they expelled Calvin and Farel, but three years
later they reversed their decision and invited them back. That
is why I chose Calvinism as a natural experiment for my first
case study. I could examine how a single human society—the
city of Geneva in the 1500s—functioned in both the presence and absence of a new religion.

Although historians of the period might disagree about
the details (all subjects have a frontier of controversy or they
wouldn't be worth studying), there is little doubt that Calvinism was instrumental in solving the problem of factionalism and helping the city of Geneva survive as a social entity,
as summarized by the eminent religious scholar Alister
McGrath:

> Events in the absence of Farel and Calvin had demon-
> strated the close interdependence of reformation and
> autonomy, of morals and morale. Although the city
> council was concerned primarily with the independence
> and morale of the city, the fact that Farel's religious
> agenda could not be evaded gradually dawned. The pro-
> Farel party probably had little enthusiasm for religious
> reformation or the enforcement of public morals; never-
> theless, it seemed that the survival of the Genevan re-
> public hinged upon them.

Try to imagine my excitement upon reading this and similar passages for the first time. Calvinism could have fit any of the five major evolutionary hypotheses outlined above, or even none of them if the entire evolutionary framework is inappropriate for the study of religion. I could have discovered that Calvin and his cronies were secretly feathering their own nests, that the religion was a parasitic culture that made everyone suffer, that it might have made sense for a hunter-gatherer group but not for a city, that it was the costly side of a coin with a beneficial nonreligious side. Instead, I discovered that Calvinism was essential for the viability of the city. Moreover, this was not my own esoteric spin but the sober assessment of the historians who best knew the facts of the matter. I especially loved the juxtaposition of the two words "morals" and "morale," including their etymological connection. The word "moral" refers to a sense of right and wrong. The word "morale" refers to motivation for action. McGrath was suggesting that for the citizens of Geneva to have a strong morale, they must have a strong and unified sense of right and wrong. Like the much smaller hunter-gatherer groups described by Chris Boehm, they needed to be a *moral community*. That was evidently what the religion provided and that even a strong, democratically elected government lacked.

Exactly how did the religion of Calvinism work its magic on the city of Geneva? To learn more, I decided to study the catechism and the Ecclesiastical Ordinances that Calvin wrote and insisted that the city adopt as a condition for his involvement. I reasoned that if anything qualifies as a "cultural genome," transmitting the essentials of a religion from person to person, it would be these. I discovered that they included not only general behavioral prescriptions such as the Ten Commandments but more specific prescriptions tailored to the Genevan social environment. The conception of God and his relationship to people, including the rendering of concepts such as original sin, faith, and forgiveness, appeared impressively designed to cultivate an attitude of civic compliance. Above all, the social practices specified by the

Ecclesiastical Ordinances appeared designed to prevent the all-important problem of cheating from within. Martin Bucer, another Protestant reformer and contemporary of Calvin, put it this way: "Where there is no discipline and excommunication, there is no Christian community." My translation: if you can't get rid of cheating by excluding the behavior (discipline) or the person if necessary (excommunication), you can forget about creating a cooperative society.

I was especially impressed by how the mechanisms for preventing cheating extended to the leaders in addition to the rank and file. The head of the Church was not a single individual but a group of pastors who made decisions by consensus, much like the egalitarian hunter-gatherers squatting around a campfire. When they couldn't agree, the decision-making circle was widened rather than narrowed. Calvin shared all the duties of a pastor, despite his enormous additional workload as primary architect of the religion and far-flung correspondence with reformers elsewhere. Elders who supervised sectors of the city had to be approved not only by the pastors and city council but also by residents of the sector. Double accounting methods were used to prevent the inappropriate use of charitable funds. These practical checks and balances were as much a part of the religion as its other-worldly elements. The egalitarian spirit of Calvinism is perhaps best illustrated by the duty of caring for dying plague victims. This life-threatening task was decided by lottery. Calvin was exempted from the lottery by decree of the city council because his death would have had a far greater impact than that of the other pastors on the fate of the church. When Calvin was succeeded upon his death by Theodore Beza, Beza himself successfully lobbied the city council to be included in the lottery. The second major evolutionary hypothesis—that religions are designed to benefit the leaders at the expense of the rank and file—can be rejected in the case of early Calvinism.

I do not mean to glamorize Calvinism or any other religion. It created a social and psychological infrastructure for its own members but did not extend its virtues to other religious groups. The Jewish community in Geneva had been expelled by the Catholics before their own expulsion by the Protestants. Calvin firmly believed that the Pope was the Antichrist. The degree of social control within the Church resembled the fundamentalist Islamic religions of today more closely than modern Christian religions, including numerous executions for heresy. In Calvin's Geneva, you could be fined for dancing inappropriately or jailed for gambling on Sunday. Calvinism and virtually all other religions of the period are guilty when judged in terms of modern human rights or the loftiest standard of universal brotherhood. As evolutionists, however, it is not our job to pass moral judgment on religions but to explain them as products of genetic and cultural evolution. I have already stressed in Chapter 5 that adaptations do not always correspond to what we regard as good and useful. When it comes to adaptations, be careful what you wish for. Whatever we might think about Calvinism from our own perspective, we can agree that it corresponds to the first major hypothesis outlined above (group-level adaptation) compared to the other four hypotheses. It succeeded by virtue of causing the city of Geneva to function as a unit, like a body or a beehive.

As I wandered the hills and valleys of the vast literature on religion, I discovered that Calvinism's immense practicality was not unusual. A wonderful book by anthropologist Stephen Lansing titled *Priests and Programmers* showed how an elaborate system of temples on the island of Bali is designed to coordinate rice agriculture. Judaism appeared especially amenable to evolutionary analysis. One of my favorite novelists is Isaac Bashevis Singer, who won the Nobel Prize for literature in 1978. I was thrilled to discover that his historical novel *The Slave* exactly paralleled the academic literature that I was reading about Jewish communities in Europe and elsewhere. Jacob, the main character of the novel, has an epiphany that I quote at the very beginning of *Darwin's Cathedral:* "But now at least he understood his religion: its essence was the relation between man and his fellows."

The religious scholar Elaine Pagels is already famous for her books on early Christianity, such as *The Origin of Satan* and *The Gnostic Gospels*. Have you ever wondered why the four Gospels of the New Testament are so different from each other, even though each is portrayed as a factual account of the life of Christ? Exhaustive scholarship has determined that they were written independently between thirty-five and one hundred years after the death of Jesus. According to Pagels, they differ not just because memory fades with time but because they were tailored to the needs of different local Christian communities. The Gospel according to Mark was written in the immediate aftermath of the Roman siege of Jerusalem and the destruction of the Temple in 70 CE. The Romans were no friends of the Christians or the Jews, but the first Christians viewed Jews in positions of authority as their primary enemy and disenfranchised Jews as the primary source of converts. The Roman destruction of the Temple was interpreted as God's punishment for the sins of the Jews, foretold by Christ, and establishment Jews were assigned the primary blame for his death.

For Matthew, writing only ten to twenty years after Mark, the main body of Judaism was controlled by a religious party known as the Pharisees, which was much less powerful during the time of Jesus. Nevertheless, the Pharisees became the primary enemy and target for blame for Christ's death in the Gospel according to Matthew. John was probably a member of a radically sectarian church composed of Jews in even more bitter opposition to the Jewish establishment than other Christian communities. His Gospel exceeds all of the others in portraying the struggle of the church as part of a cosmic struggle between good and evil. Ever since, according to Pagels, Christian communities that find themselves battling for their lives have found special inspiration and comfort in the Gospel according to John.

Luke is probably the only Gospel written by and for Gentiles, which is amply reflected in its contents. When Jesus preaches in his hometown of Nazareth, he is well received in the other three Gospels, but in Luke he is almost thrown off a cliff because he says that God will bring salvation to the Gentiles. According to Luke, all Jews clamor for the death of Jesus, not just the Pharisees or the Jewish establishment. Pontius Pilate, the Roman governor of Judaea, is portrayed as a reasonable man who tries his level best to save Jesus, even though nonbiblical sources describe him as a man of "inflexible, stubborn, and cruel disposition" whose administration was marked by "greed, violence, robbery, assault, abusive behavior, frequent executions without trial, and endless savage ferocity." When Jesus dies upon the cross, Luke has a Roman centurion exclaim, "Certainly this man is innocent!" (23:47).

Even more interesting are the Gospels that didn't make it into the New Testament. When the Christian movement gained sufficient momentum, it became necessary to impose uniformity by canonizing some of the religious teachings and denouncing the rest. During the late second century, bishops of the Christian church that by now was calling itself orthodox met to assemble the New Testament and brand everything else, in the words of one of the bishops, "an abyss of madness, and blasphemy against Christ." Fortunately, some of the banned documents survived and have been retrieved from the abyss by scholars such as Elaine Pagels. One is the Gospel according to Thomas, which encourages the believer to embark upon a journey of self-discovery rather than conforming within a close-knit group. According to Pagels, the Gospels that made it into the New Testament were chosen for the following reason:

> The author of Mark, then, offers a rudimentary model for Christian community life. The gospels that the majority of Christians adopted all follow, to some extent, Mark's example. Successive generations found in the New Testament gospels what they did not find in many other elements of the early Jesus tradition—a practical design of Christian communities.

Pagels doesn't use the *E*-word, but she is unmistakably describing a process of cultural evolution. Versions of Christianity that build strong communities survive while other versions fall apart. The elements of a religion required for survival depend upon the surrounding social environment, so religions necessarily diversify as they evolve. I never imagined at the beginning of my inquiry that the New Testament might count as a fossil record of local cultural adaptation! As with Calvinism, this was not my own esoteric interpretation but the opinion of respected scholars.

Even religious leaders, and not just religious scholars, sometimes describe religion in evolutionary terms without using the *E*-word. Consider the following passage by John Wesley, who founded the variety of Christianity known as Methodism:

> I do not see how it is possible, in the nature of things, for any revival of religion to continue for long. For religion must necessarily produce both industry and frugality. And these cannot but produce riches. But as riches increase, so will pride, anger, and love of the world in all its branches.

Wesley is saying that religions are so good at providing practical benefits that their members become wealthy, whereupon they lose the incentive to cooperate and try to loosen the very strictures that lifted them out of poverty. In addition, religions are *not* entirely fair in practice. Some members do gain at the expense of others (the second major hypothesis outlined above), prompting the have-nots to leave and create their own "purified" church. Religions not only adapt to their social environments but also *change* their social environments, leading to an endless cycle of corruption and renewal that has been documented by scholars for all religious traditions, around the world and throughout history.

What about forgiveness, the specific topic that I was funded by the Templeton Foundation to study? Christian forgiveness is often summarized by the phrase "turn the other cheek," but this single behavioral prescription is far too simple from an evolutionary perspective. Any religion worth its salt must provide many rules about forgiveness that are employed flexibly, depending upon the situation. The first requirement is to define the group and isolate it from the rest of society so that in-group and out-group behaviors can be regulated separately. Anyone familiar with the gospels knows that Jesus demanded total commitment, eclipsing not only one's prior religion but also one's immediate family. In the parable of the talents (Luke 19:12–27), Jesus tells a story of a nobleman who travels to a distant land "to get royal power for himself." Upon his return, he demands the death of those who were disloyal in his absence: "As for those enemies of mine who did not want me to be king over them—bring them here and slaughter them in my presence." This facet of Christianity is the polar opposite of "turn the other cheek"— *and it must be* to perform its function. Before there can be a strongly committed group, there must be perceived (or actual) dire consequences for leaving the group. Fast-forwarding to the sixteenth century, Calvin's God metes out forgiveness with numerical precision: those who enter and then leave the faith are cursed to the fourth generation, those who keep the faith are blessed for a thousand generations, and so on.

Within the group, members do not "turn the other cheek" with respect to social transgressions but follow a detailed set of rules, as described by the same member of the Hutterite faith who initially piqued my interest in religion by describing his group as like a body and a beehive:

> The bond of love is kept pure and intact by the correction of the Holy Spirit. People who are burdened with vices that spread and corrupt can have no part in it. This harmonious fellowship excludes any who are not part of the unanimous spirit.... If a man hardens himself in rebellion, the extreme step of separation is unavoidable. Otherwise the whole community would be dragged into his sin and become party to it.... The Apostle Paul therefore says, "Drive out the wicked person from among you."
>
> In the case of minor transgressions, this discipline consists of simple brotherly admonition. If anyone has acted wrongly toward another but has not committed a gross sin, a rebuke and warning is enough. But if a brother or a sister obstinately resists brotherly correction and helpful advice, then even these relatively small things have to be brought openly before the Church. If that brother is ready to listen to the Church and allow

himself to be set straight, the right way to deal with the situation will be shown. Everything will be cleared up. But if he persists in his stubbornness and refuses to listen even to the Church, then there is only one answer in this situation, and that is to cut him off and exclude him. It is better for someone with a heart full of poison to be cut off than for the entire Church to be brought into confusion or blemished.

The whole aim of this order of discipline, however, is not exclusion but a change of heart. It is not applied for a brother's ruin, even when he has fallen into flagrant sin, into besmirching sins of impurity, which make him deeply guilty before God. For the sake of example and warning, the truth must in this case be declared openly and brought to light before the Church. Even then such a brother should hold on to his hope and his faith. He should not go away and leave everything but should accept and bear what is put upon him by the Church. He should earnestly repent, no matter how many tears it may cost him or how much suffering it may involve. At the right time, when he is repentant, those who are united in the Church pray for him, and all of Heaven rejoices with them. After he has shown genuine repentance, he is received back with great joy in a meeting of the whole Church. They unanimously intercede for him that his sins need never be thought of again but are forgiven and removed forever.

I have quoted this long passage to show how a religion can embody detailed instructions for how to measure out forgiveness in particular situations. If you are familiar with the branch of mathematics called game theory, you will immediately recognize the elements of this religious passage as similar to the elements of game theoretic strategies that promote the evolution of cooperation—retaliation leavened with just the right amount of generosity, contrition, and forgiveness conditioned upon a change of behavior.

What role does "turn the other cheek" play in this complex system of if-then rules? Jack Miles, the Pulitzer prize–winning author of *God: A Biography*, provides part of the answer in an essay titled "The Disarmament of God." According to Miles, the Hebrew God was essentially a warrior who commanded his people to fight and promised them victory in the future, no matter how many defeats they had endured in the past. The Christian God reflected the reality that military victory was no longer possible and the only strategy for survival involved a more peaceful coexistence. The Christian God laid down his arms. This was such a radically different social strategy that the Christian God could be said to be a different God entirely from the Hebrew God, as some scholars have noted. Cultural evolution seldom involves such radical discontinuities, however, so Christians imagined their God as continuous with the past. In any case, "turn the other cheek" can be a successful nonmilitaristic strategy in between-group competition, as subsequent events amply confirm. Once Christians became politically powerful, cultural evolution promoted the resumption of militaristic strategies, as in the Crusades.

By the end of my three-year journey, I was convinced that religious believers are essentially correct when they describe their groups as like bodies and beehives. Evolution is a messy and multifactorial process. All of the five major evolutionary hypotheses might have a degree of validity, but if you could say only one thing about religion, it would be close to Durkheim's definition quoted earlier in this chapter, which requires a process of cultural group selection to explain in modern evolutionary terms. My journey was haphazard, however, which is why I gathered my unsuspecting band of students to help me sample religions at random, like the blindfolded impartial Goddess of Justice. I wish I could tell you that my students became a crack team of evolutionary religious scholars in the space of a single semester. It would be a great plot for a made-for-TV inspirational movie. Alas, they failed to achieve the reliability standards required for the part of the study that was designed to convert descriptive information into numerical form. They did write some very informative essays with hundreds of references to books and articles on the religions in the sample, however, enabling me to continue the survey on my own after the semester was over. The results have been published in a paper titled "Testing Major Evolutionary Hypotheses About Religion with a Random Sample," which you can download from my Web site. You are amply prepared to enjoy and evaluate it, despite the fact that it is written for a professional audience. Suffice it to say that the random sample confirms the conclusions that I reached in *Darwin's Cathedral* based on my more haphazard journey. The beauty of random sampling is that, barring a freak sampling accident, valid conclusions about the sample are also valid for *all* of the religions in the sixteen-volume *Encyclopedia of World Religions*, from which the sample was drawn.

Since writing *Darwin's Cathedral* I have traveled the world speaking about evolution and religion to audiences of all sorts. I end my talk with the following passage from Darwin's autobiography about a field trip that he took as a young man with his professor Adam Sedgwick to a valley in Wales.

We spent many hours in Cwm Idwal, examining all of the rocks with extreme care, as Sedgwick was anxious to find fossils in them; but neither of us saw a trace of the wonderful glacial phenomena all around us; we did not notice the plainly scored rocks, the perched boulders, the lateral and terminal moraines. Yet these phenomena are so conspicuous that ... a house burnt down by fire did not tell its story more plainly than did this valley. If it had still been filled by a glacier, the phenomena would have been less distinct than they now are.

This passage wonderfully illustrates *the need for a theory to see what is in front of our faces*. Darwin and Sedgwick could not see the evidence for glaciers because the theory of glaciation had not yet been proposed. With the theory in mind, the confirming evidence became so obvious that the glaciers might as well have still been present. Detailed measurements, statistics, and other trappings of modern science were unnecessary. Descriptive information carefully gathered by the geologists of the day was sufficient.

Darwin's theory of natural selection represents another transformation of the obvious. It was established entirely on the basis of descriptive information about the living world, carefully gathered by the naturalists of the day, most of whom thought they were studying God's handiwork. Creationism could not make sense of the facts, any more than Darwin and Sedgwick could find fossils in the valley of Cwm Idwal, but with the right theory in mind the confirming evidence became so obvious that we might as well have been present during the origin and diversification of species.

I think that a similar transformation of the obvious will take place for our understanding of religion. Religious scholars are like the geologists and naturalists of Darwin's day. Over the decades, they have created mountains of carefully gathered descriptive information about religions around the world and throughout history. I know, because I have spent years wandering those mountains without coming to the end of them. Some religious scholars are religious in their personal convictions, others are guided by one theoretical framework or another, but no one has even remotely made sense of it all. Fortunately, the future need not be like the past, as we have seen in the case of geology and biology. With the right theory, the confirming evidence can become so obvious that we might as well have been present during the origin and diversification of religions, as indeed we are, since the process continues to take place all around us. It doesn't matter that the information is descriptive and lacks the trappings of modern science. Quantification *refines* but does not *define* the scientific process, which is why Darwin was able to establish his theory on the strength of purely descriptive information.

Evolution and religion can no longer occupy opposite corners of human thought. Evolutionists must include religion as part of what it means to be human, and when religious believers describe their groups as like bodies and beehives, they have nothing to fear from science.

Is There Anyone Out There?
Is There Anyone Up There?

I'M NOT SOPHISTICATED ABOUT music, but I know what I like, and that includes Ray Charles. I don't remember how I discovered him, but while my friends were singing the Beatles and the Supremes, I was belting out renditions of "What'd I Say" and "You Don't Know Me." I recently splurged on a five-CD collection that includes newer songs in addition to the old classics. One of my favorites is a mighty gospel number called "Is There Anyone Out There?" Just because I'm an evolutionist doesn't mean that I can't be moved by gospel music, and when the raw emotionality of Ray's voice is joined by what sounds like a choir of a thousand angels, I get the same electric feeling as your average religious believer.

But wait! The title of the song is "Is There Anyone *Out* There?" not "Is There Anyone *Up* There?" As I listen carefully to the words, I discover that the song is entirely about the need for social connections. Aside from one or two Lord-have-mercies, there isn't a single reference to a higher power, but that does not detract from the power of the song and its underlying message.

Religions can be immensely effective at forming social connections, as we have just seen, providing an answer to Ray's question "Is there anyone out there?" But religion is inherently about the second question also, even if it was absent from this particular song. Religious scholars even use the terms "horizontal" and "vertical" in ways that correspond to the "out" and "up" of my two questions, as in the following definition of Islam from Eliade's encyclopedia:

> A noun derived from the verb aslama ("to submit or surrender [to God]"), designates the act by which an individual recognizes his or her relationship to the divine and, at the same time, the community of all of those who respond in submission. It describes, therefore, both the singular vertical relationship between the human being and God and the collective, horizontal relationship of all who join together in common faith and practice.

Why do religions include a vertical dimension in addition to a horizontal one? Why should anyone care about the existence of a higher power, as long as there are people out there to help in one's hours of need? Evolutionary theory might provide answers to these questions with two words of its own: proximate and ultimate.

All adaptations require two explanations, as I described in Chapters 7 and 10. Why do flowers bloom in spring? The ultimate explanation, based on survival and reproduction, is because spring is the best time to bloom. Perhaps flowers that bloomed earlier were nipped by frost, while those that bloomed later didn't have time to develop their fruits. The proximate explanation is based on the physical mechanisms that actually cause flowers to bloom in spring, such as sensitivity to day length. Notice that day length by itself has no effect on survival and reproduction. It is merely a signal that reliably causes flowers to bloom at the best time with respect to other environmental factors. In general, the proximate explanation for a trait need bear no relationship whatsoever to the corresponding ultimate explanation, as long as it reliably produces the trait that survives and reproduces better than other traits.

Returning to religion, a given belief or practice might exist because it enhances survival and reproduction—for example, by causing the group to function well compared to other groups—but this is only the ultimate explanation. A complementary proximate explanation is needed that need bear no relationship to the ultimate explanation, other than to reliably cause the trait to occur. Perhaps a religious believer helps others because she wants to help others, or perhaps because she wants to serve a perfect God who commands her to help others. If these two proximate mechanisms are equally effective at motivating helping behavior, evolution will be indifferent as to which one evolves. If wanting to serve a perfect God is more powerfully motivating than directly wanting to help others, then it is likely to evolve as the proximate mechanism, even though it is less obviously related to the behavior that it produces and requires belief in an agent for which there is no tangible proof.

In short, the proximate/ultimate distinction provided by evolutionary theory might go a long way toward explaining the vertical and horizontal dimensions of religion. If you think that this observation is elementary, it is only because you are thinking as an evolutionist. In the past, religion has baffled the scientific imagination because it appears so *irrational*. How can anyone entertain such outlandish beliefs for which there is no tangible proof whatsoever? These questions assume that rational thought and tangible proof are appropriate gold standards for judging religion, but why should this be so? From an evolutionary perspective, there is only one gold standard for judging religious thought, rational thought, or any other form of thought—*what does it cause people to do?* Once we employ the right gold standard, it becomes obvious that rational thought is capable of producing successful behaviors only under some conditions and that departing from rational thought can be highly successful under other conditions. The irrational nature of religious thought ceases to become baffling to the scientific imagination, but an ounce of evolutionary thinking was required to provide the right gold standard.

Since writing *Darwin's Cathedral*, I have spoken with many religious believers who feel that my focus on practical benefits misses the essence of the religious experience, which is a deeply felt relationship with God. I agree with them as far as the *psychological* religious experience is concerned, but that is exactly what the proximate/ultimate distinction leads us to expect. I could be right that religion is all about practical benefits in terms of what religious belief causes people to do (the ultimate explanation, which corresponds to the horizontal dimension of religion), and they could be right that their own religious experience is based far more on their relationship with God than on practical benefits for themselves or anyone else (the proximate explanation, which corresponds to the vertical dimension of religion). The proximate explanation need not bear any relationship to the ultimate explanation other than reliably causing the right behavior, as we have seen with our example of the flowering plant. By the same token, people fall in love in part to have children (an ultimate explanation), but that doesn't remotely describe the subjective experience of falling in love (the proximate explanation).

Religious beliefs are so diverse that they seem to defy categorization. Supernatural agents play a much larger role in some religions than others. The Buddha refused to be associated with any gods; he said that he was merely awake and had found a path to enlightenment. It's true that there are gods associated with Buddhism as a major religious tradition, but they seem to exist at the edges rather than the center. Confucianism is even more frankly pragmatic than Buddhism, which does not detract from its power or historical influence. Even religions that rely upon supernatural agents cannot be defined entirely in terms of such beliefs, or else there would be no way to distinguish between God and the Easter Bunny.

Evolutionary theory does a good job explaining biological diversity and can go a long way toward explaining patterns of religious diversity. In the first place, both biological species and varieties of religion are much *less* diverse than they could be because we see primarily the survivors. Bears, badgers, and buffalo are highly adapted to their respective ecological niches. If they could mate with each other or if we could produce hybrids by some dubious feat of genetic engineering, the intermediate forms would be quickly eliminated by natural selection because they don't have the right combination of traits to survive and reproduce. That is why bears, badgers, and buffalo don't mate with each other, or even with much more closely related species! Adaptive diversity is great but still a tiny fraction of nonadaptive diversity.

If most religions are as adaptive at the group level as I claim in *Darwin's Cathedral*, then the vertical dimension of religion must be very tightly linked to the horizontal dimension. It is easy to imagine religious beliefs that inspire people to do all sorts of things—to help cheaters, to commit suicide, to care only about oneself. These beliefs *could* exist but evolutionary theory predicts that they *won't* or, rather, that they are greatly underrepresented, because they do not lead to sustainable communities.

In my random sample of religions, I was amazed at how tightly the vertical and horizontal dimensions were yoked to each other. Joseph Smith translated the Book of Mormon by putting his face into a hat containing magic stones. The Cao Dai religion in Vietnam began with the Supreme Being revealing himself through a Ouija board. Saint Catherine of Sienna had a vision of Christ at the age of six and took a vow of virginity against her family's wishes. Nahman of Bratislava (a Jewish spiritual leader) locked himself in his parents' attic for long periods of time in an attempt to gain nearness to God. His disapproval of secular desires went so far that he didn't want a following, which only enhanced his reputation in the eyes of his followers. Jain ascetics do everything possible to avoid killing the minutest of life-forms. Those who are not religious can be forgiven for regarding these people and their followers as crazy. Remarkably, however, these and other "crazy" religious beliefs lead to practices that are the very opposite of crazy!

What accounts for the linkage between the frequently crazy-appearing vertical dimension and the immensely practical horizontal dimension of religion? Part of the answer involves variation and selection, pure and simple. New religions are created all the time and the vast majority fall apart without attracting any notice. Then there are the spectacular religious failures, such as the mass suicide of the People's Temple cult in 1978. Many religious activists flocked to Geneva to learn the secrets of Calvinism, but few flocked to Guyana to learn the secrets of Jim Jones. To some extent, religious beliefs are just like genetic mutations: they arise arbitrarily and only the ones that work are retained by imitation and selection.

This is only part of the story, however. The human mind does much more than cast about randomly for solutions. Rational choice is a largely conscious procedure for seeking solutions to life's problems and there are many other procedures that operate beneath conscious awareness, as we saw in Chapters 26 and 27. These mental adaptations evolved by a raw process of variation and selection in the past but themselves are highly strategic. It is likely that we are psychologically adapted to create beliefs with practical objectives in mind. In other words, the vertical and horizontal dimensions of religion are already linked by our mental processes, even though we might not be consciously aware of the linkage. When Joseph Smith was pushing his face into the hat with the magic stones and the originators of the Cao Dai religion were gathered around the Ouija board, they probably had something in mind, whether they knew it or not, that made their incipient religions more than just a roll of the dice.

Regardless of how the vertical and horizontal dimensions of religion become yoked to each other, most enduring religious beliefs lead to practical benefits, or so I claim based on *Darwin's Cathedral* and my random sample. Religious diversity is much less than it could be because it is restricted to islands of functionality, just like biological diversity. Even so, any given adaptive trait can be caused by many different proximate mechanisms. To continue my biological example, not all plants that bloom in spring are sensitive to day length. Some species are sensitive to temperature, others have an internal biological clock, and so on. There are many ways to skin a cat, and the particular proximate mechanism that evolves in a particular species can be a matter of chance or historical contingency. In just the same way, a single behavior associated with religion can be motivated by many different beliefs and practices. Suppose that you lose your wallet, which is stuffed with credit cards and a thousand dollars in cash. Miraculously, it is returned to you with all of its contents. What prompted such a noble deed? Perhaps the finder acted out of a sense of duty, or because it made him feel good, or because he regarded it as a ticket to heaven, or because he was with others who shamed him into it. You don't know and there's a sense in which you shouldn't care, as long as you got your wallet back. It is quite possible that religions are similar in the behaviors that they promote (such as returning a wallet) but differ in how they get the job done.

Let's see how these ideas can be used to study an important religious concept such as the afterlife. One of the most common theories of religion at the level of cocktail party conversation is based on our fear of death. According to this theory, self-awareness evolved in our species by producing many practical benefits, but it had the unfortunate side effect of making us aware of our own demise. The concept of a glorious afterlife was invented to allay our fear of death. You must have heard this theory dozens of times. Probably no one used the E-word, but it is easily classified as a by-product theory, similar in spirit, if not in detail, to the more formal theories of Stark, Bainbridge, Boyer, and Atran mentioned briefly in Chapter 28. It's a perfectly plausible scenario, but it doesn't fit the facts. It predicts that belief in a glorious afterlife should be a religious universal, which isn't the case. There's nothing comforting about the Greek afterlife. Even Judaism places far less emphasis on the afterlife than Christianity does. Historically, Jews already had a strong motivation to establish the nation of Israel on earth. In general, all enduring religions must highly motivate their members, but the specific means by which they do so can be highly variable. This is expected from the many-to-one relationship between proximate mechanisms and any single behavior. Even a belief as prominent in some religions as a glorious afterlife can be peripheral or absent in others.

Religion is in part a subject like any other, a particular feature of a particular species that needs to be explained, like infanticide in beetles or shyness in sunfish. Religion is especially fascinating from this purely intellectual perspective because it is such unexplored territory, at least as far as evolutionary theory is concerned. The vast majority of scholars and social scientists who study religion know much less about evolution than you do on the basis of this book. I therefore feel the excitement of discovery, similar to how the early naturalists must have felt when they beheld a tropical rain forest for the first time.

Of course, religion is not a purely intellectual subject, especially for religious believers. I might get excited by the prospect of heaven as one of many proximate mechanisms for motivating behavior, but someone who fervently believes in heaven is likely to react differently. Can my tall claim about bringing evolution and religion harmoniously together possibly be fulfilled for the religious believer in addition to the scientist studying religion?

To the religious believer's dilemma, I need to add a dilemma of my own. In *Darwin's Cathedral* I make a distinction between factual and practical realism. A belief is factually realistic when it accurately describes features of the world. A belief is practically realistic when it helps people survive and reproduce in the world. As a compassionate human being, I have the same strong interest in practical realism as any religious believer. When I see religions provide answers to Ray's question "Is there anyone out there?" I feel nothing but admiration. As a scientist, I also have a strong commitment to factual realism, which prevents me from accepting many religious answers to the question "Is there anyone up there?" My ideal religion would provide a strong horizontal dimension *and* a strong vertical dimension that is fully consistent with scientific knowledge. Is such a religion possible, or can the vertical dimension only become strong by departing from factual realism? That is my dilemma, and I feel it as keenly as a religious believer confronting the possibility that her particular religious beliefs might "merely" be practically and not factually realistic.

I do not have an answer to my dilemma, much less one that can be validated scientifically, but slowly and after much thought I am gaining a conviction that I can have my cake and eat it too. In the first place, many religions are constructed to be consistent with the factual knowledge of their day and become factually unrealistic only after the passage of time. The ancient Hebrews didn't know that light came from the sun, which made it reasonable for them to say that God created light before the sun. When I read Calvin's catechism, I was impressed by its *scientific* validity, based on knowledge available at the time. It is important to remember that before Darwin, there was no way to explain the immense functionality of nature and human affairs without invoking a deity. Religions might not easily keep pace with changes in factual knowledge, especially because religious beliefs are so often presented as eternal, but they do not always set out to flaunt factual knowledge.

In the second place, major religious traditions such as Buddhism and Confucianism come close to my ideal religion by providing strong horizontal dimensions with vertical dimensions that adhere closely to factual realism. His Holiness the Dalai Lama has just written a book on this very topic, titled *The Universe in a Single Atom: The Convergence of Science and Spirituality*. Everyone should read this book to learn how science does not conflict with religion per se, only with certain religious traditions. In Buddhist thought, "to defy the authority of empirical evidence is to disqualify oneself as someone worthy of critical engagement in a dialogue." The Dalai Lama became curious about science as a little boy, staring at the moon through a telescope that belonged to the thirteenth Dalai Lama. His political exile made him a worldwide spiritual ambassador and also enabled him to satisfy his curiosity about science. He has made an earnest effort to learn about the most recent theories of cosmology, physics, evolution, psychology, and neurobiology, through a worldwide network of advisors and periodic "Mind and Life" conferences that he organizes.

His humility is wonderfully refreshing. With respect to Einstein's theory of relativity, he remarks: "Appreciating the full nature of the twins paradox involves understanding a set of complex calculations which I am afraid are beyond me." He cheerfully concludes that "Buddhism ... must be willing to adapt the rudimentary physics of its early atomic theories, despite their long-established authority within the tradition." When he discovered that neurobiologists were interested in studying the brain states associated with meditation, he even put them in touch with some hermits that he knew!

> For the hermits who have chosen a life of solitude in the mountains ... such experimentation constitutes a profound intrusion into their lives and spiritual practice. It is not surprising that initially many were reluctant. Apart from anything else, most simply couldn't see the point, other than satisfying the curiosity of some odd men carrying machines. However, I felt very strongly (and still feel) that the application of science to understanding the consciousness of meditators is most important, and I made a great effort to persuade the hermits to allow the experiments to take place. I argued that they should undergo the experiments out of altruism; if the good effects of quieting the mind and cultivating wholesome mental states can be demonstrated scientifically, this may have beneficial results for others. I only hope I was not too heavy-handed. A number of hermits accepted, persuaded, I hope, by my argument rather than simply submitting to the authority of the Dalai Lama's office.

How can a religious leader such as the Dalai Lama, whose status among Tibetan Buddhists is equivalent to that of the Pope among Catholics, be so unthreatened by science? As he puts it, "The greatness of the Buddha as a spiritual teacher lies not so much in his mastery of various fields of knowledge as in having attained the perfection of boundless compassion for all beings." Elsewhere he states, "The principal aim of Buddhist psychology is not to catalog the mind's makeup or even to describe how the mind functions; rather, its fundamental concern is to overcome suffering, especially psychological and emotional afflictions, and to clear those afflictions." In this and other passages, *he demonstrates an explicit awareness that the ultimate purpose of religion is to create a strong horizontal dimension, with the vertical dimension subservient to that goal.* If I were ever lucky enough to converse with the Dalai Lama, I think that we would be largely in agreement.

Other religious traditions might not be as explicit about the ultimate purpose of religion and the vertical dimension's subservient role, but it is there just beneath the surface. I am not only willing but eager to talk with religious believers, and I have had many opportunities since writing *Darwin's Cathedral.* I have spoken to book clubs in Baptist churches, spent an idyllic week at a Unitarian retreat on an island off the coast of Maine, and engaged in a televised conversation with Benedictine monks and faculty from St. John's University in Minnesota. The Benedictine order is one of the oldest surviving social organizations in human history, as I learned in preparation for my visit, and the monastery where I was graciously housed is the oldest in North America. The Benedictine Rule (not *rules,* as I was informed in no uncertain terms) could have easily been included in *Darwin's Cathedral* alongside Calvin's catechism and Ecclesiastical Ordinances as a blueprint for community life. The entire north face of the St. John's Abbey Church is a stained-glass window in a honeycomb motif, making me feel right at home.

My encounters across the science-religion divide are invariably cordial, not because we are on our best behavior, but because we have so much in common. There is so much more to religion than belief in supernatural agents—building strong communities, helping others in need and being helped in return, transmitting our best values to our children, and the possibility of transformative change. These make up the horizontal dimension of religion and can be largely affirmed by evolutionary theory, or at least by the first major hypothesis outlined in Chapter 28. If you are a religious believer and could choose among the five major evolutionary hypotheses, which would you prefer? It's true that religions are designed primarily to take care of their own members, but are secular groups any different? In *Darwin's Cathedral* I say that criticizing religions for failing to achieve universal brotherhood is like criticizing birds for failing to break the sound barrier. Can't we admire them for flying as well as they do? As for the mayhem created by interactions among religious groups, according to my random sample of religions, the proportion that spread by violent conflict is actually rather small. Most religions spread by offering a good deal to their members. What is the track record of nonreligious groups, such as communist and totalitarian regimes, by comparison?

In my reading and personal encounters, I am heartened by the number of people who see their religions clearly and remain strong in their faith without requiring departures from factual reality. Consider Myles Horton, one of the great social activists of the twentieth century. He was born into a Calvinist family in Savannah, Tennessee, not far from the birthplace of Sir John Templeton and Dayton, the site of the Scopes trial. His mother was a pillar of her church, the kind of woman whom everyone in the community leaned upon for advice and support. As soon as little Myles was old enough to read the catechism, he went to his mother for advice. Here is how he recalls the event in his autobiography *The Long Haul:*

> One day I went to my mother and said, "I don't know, this predestination doesn't make any sense to me, I don't believe any of this. I guess I shouldn't be in this church." Mom laughed and said, "Don't bother about that, that's not important, that's just preacher's talk. The only thing that's important is that you've got to love your neighbor." She didn't say "Love God," she said "Love your neighbor, that's all it's all about."... It was a good non-doctrinaire background, and it gave me a sense of what was right and what was wrong.

Horton founded the Highlander Folk School in 1932, which taught leadership and organizational skills to both blacks and whites in defiance of segregation laws. According to social commentator Bill Moyers, "He's been beaten up, locked up, put upon and railed against by racists, toughs, demagogues, and governors." He did not require complicated beliefs that depart from factual reality and neither did his mother.

Benjamin Franklin expressed a similar pragmatic view of religion in his autobiography:

I grew convinced that truth, sincerity, and integrity in dealings between man and man were of the utmost importance to the felicity of life. . . . Revelation had indeed no weight with me, as such; but I entertained an opinion that, though certain actions might not be bad *because* they were forbidden by it, or good *because* it commanded them, yet probably these actions might be forbidden *because* they were bad for us, or commanded *because* they were beneficial to us, in their own natures, all of the circumstances of things considered.

For Myles Horton and Benjamin Franklin, understanding the practical wisdom of religion strengthened rather than weakened their sense of right and wrong.

I'm not in the same league as these men but I do have an anecdote to share from my own life. When our two daughters were small, we taught them about keeping promises and telling the truth in a way that was completely practical. What could be better than to know with certainty that others are telling you the truth and keeping their promises? All you need to do is play by the same rules. With our younger daughter, Tamar, I illustrated by giving her a magic phrase that she could say anytime she wanted me to stop doing something: "I really, really don't want you to do this!" I would illustrate by tickling her, and when she uttered the magic phrase it was like being protected by an invisible shield. None of this was very deliberate but it worked better than I could have imagined. Our kids were so scrupulous that they didn't even resort to crossing their fingers. Each in turn solemnly announced to their teachers in elementary school that they couldn't repeat the part about God during the pledge of allegiance, and I *promise* that we didn't put them up to it! Our advice would have been to keep quiet, but that was beneath their moral scruples. Their little friends were aghast and warned about going to hell, but amazingly, their religious upbringing made them much less scrupulous about keeping promises and telling the truth (at least at school) than our daughters!

Our family social convention worked only because we

kept up our end of the bargain. Never once did I continue tickling Tamar or perform any other objectionable act after she uttered her magic phrase. If I did, there would have been hell to pay, and Tamar is capable of inflicting it, I can assure you. The social convention was *sacred*, you see, and we respected it accordingly. My own experience, along with examples such as the Dalai Lama, Myles Horton, and Benjamin Franklin, is slowly giving me faith that the blessings of religion do not require departures from factual reality.

30

Ayn Rand: Religious Zealot

RELIGIONS HAVE A WAY of departing from factual realism in their drive toward practical realism, as I have shown in the last two chapters. The departures are by no means restricted to belief in supernatural agents. As Elaine Pagels shows in her analysis of the Christian gospels, people and events are also freely invented and altered. When the only purpose of a belief system is to provide a blueprint for action, any and all aspects of factual reality are fair game.

As soon as we begin thinking this way about religion, it becomes obvious that we must extend our analysis beyond religion. Patriotic histories of nations have the same distorted and purpose-driven quality as religions, a fact that becomes obvious as soon as we consider the histories of nations other than our own. Intellectual movements such as feminism and postmodernism are often shamelessly open about yoking acceptable truths to perceived consequences. That's what it means to be politically correct. Scientific theories are not immune. Many scientific theories of the past become weirdly implausible with the passage of time, just like religions. When this happens, they are often revealed as not just wrong but as purpose-driven. My favorite examples concern scientific theories that support conventional beliefs about the role of women in society. In the nineteenth century, it was received scientific wisdom that going to college would interfere with ovarian development. As late as the 1970s, women were barred from running marathons because it was thought that running more than a few miles would damage their bodies. We can laugh at these theories today as just plain wrong, but only because they were rejected by factual evidence—women running marathons and graduating from college with functioning ovaries. Scientific theories cannot be expected to approximate factual reality when they are proposed, but only after they have been winnowed.

These and other belief systems are not classified as religions because they don't invoke supernatural agents, but they are just like religions when evaluated in terms of factual and practical realism. In all cases, they function primarily as blueprints for action and depart from factual realism along the way. The presence or absence of supernatural agents—a particular departure from factual realism—is just a detail.

It is humbling to contemplate that the concerns typically voiced about religion need to be extended to virtually all forms of human thought. If anything, nonreligious belief systems are a greater cause for concern because they do a better job of masquerading as factual reality. Call them stealth religions.

A good example of a stealth religion is Ayn Rand's philosophy of objectivism. If you are familiar with Ayn Rand and her books, you will appreciate the irony of my chapter title. She was an ardent atheist and her philosophy is founded upon the virtue of rational thought. When people labeled her an individualist, she corrected them by calling herself a rationalist. How on earth can I call her a religious zealot?

I did not set out to study Ayn Rand but discovered her as part of a project on language. Some words have very precise definitions, such as the parts of a ship. It's easy to understand why the parts of a ship are precisely defined, because it is important to refer to them unambiguously. When there is a great storm and everyone's life is in danger, you can't say, "Take in that rope over there! No! Not that one! *That* one!"

Other words seem lost in a fog of ambiguity. Take the word "selfish"; it is based on a nebulous mix of actions and motives operating over the short and long terms. To some it is a way not to behave, while to others it is a principle that explains all rational behavior. Why isn't the word "selfish" defined as precisely as the word "halyard"? Nobody with whom I talked seemed to know or even to have thought of the question. One philosopher merely shrugged and replied that people have agreed to disagree.

I thought that I had a novel answer to this question based on an analogy with natural ecosystems, whose many species coexist by surviving and reproducing in different ways. People are so behaviorally diverse that they are more like an ecosystem than a single species, as I have stressed in previous chapters. The human ecosystem includes many belief systems rather than many species, each instructing its members to behave in different ways. Within a given belief system, a word such as "selfish" might have a consistent meaning, like the word "halyard," but consistency *across* belief systems should not be expected. After all, belief systems must often use the same words in different ways if they are to motivate different behaviors.

To test my idea, I decided to collect and classify uses of the word "selfish" and related words, much as an entomologist would collect and classify species of butterfly. I developed a checklist of "traits" that are associated with the word "selfish" at least some of the time. When a behavior is described as selfish, what are its effects on self and others? Are the effects short-term or long-term? Are they material or psychological? Whenever I encountered the word "selfish" in my reading, I would grab it like a butterfly in my net and classify it by filling out my checklist. For example, here is a passage from a letter that William James wrote to his mother as a young man asking for money: "When you speak of your own increased expenses, I feel very guilty and *selfish* in entertaining any projects which look in the least like extravagance." The behavior labeled as selfish (asking for money) is good for the self (James), bad for the other (his mother), there is no distinction between short-term and long-term, and the effects are primarily material (money). The fact that James felt guilty (a negative psychological effect) did not alter the status of the behavior as selfish in his own mind.

This meaning of the word "selfish" is so common that we tend to take it for granted, but now let's see how the same word is used in a *Reader's Digest* article titled "How Love Came Back." The author admits to himself that he has been a *selfish* husband (there is the word usage that I grab with my net). His selfishness is expressed mostly in small ways, like chiding her for being late and insisting on watching his preferred TV shows. He decides to be really nice to her on their vacation and discovers that he also becomes happier. His wife is so dumbfounded by his kindness that she decides she must have a terminal disease. The story ends with him gathering her into his arms, saying, "No, honey! You're not dying; I'm just starting to live!"

If you have managed to suppress your gag reflex, we can proceed to classify this particular specimen of the word "selfish." It refers to a behavior (being inconsiderate) that is good for the husband and bad for his wife over the short term, but bad for both of them over the long term. The effects are primarily psychological (happiness). The most important feature of this usage that distinguishes it from the previous one is that selfishness has been turned from a win-lose proposition to a lose-lose proposition. In the tiny little world built by the tiny little story, there are no conflicts of interest.

This "species of thought," as I like to refer to it, is extremely common. In Oscar Wilde's children's story "The Selfish Giant," a giant prevents children from playing in his courtyard, only to have winter descend permanently on his domain. Spring returns when he allows the children to return. A slogan for the Red Cross states, "Do something for nothing. And you'll get everything." The word "selfish" and various synonyms are used nineteen times in the first five chapters of the Alcoholics Anonymous manual. In every case, the long-term consequences are portrayed as negative for the selfish person, as in this example: "Selfishness—self-centeredness! That, we think, is the root of our troubles. Driven by a hundred forms of fear, self-delusion, self-seeking and self-pity, we step on the toes of our fellows and they retaliate. Sometimes they hurt us, seemingly without provocation, but we invariably find that at some time in the past we have made decisions

based on self which later place us in a position to be hurt. . . .
Above everything, we alcoholics must get rid of this selfishness. We must, or it kills us!"

In the AA manual, at least, the word "selfish" has the same
consistency and precision as the word "halyard." Three observations can be made about this "species of thought." First, it is
clearly designed to accomplish a change in behavior by getting
people to be more other-oriented. Second, it might well provide good advice in practical terms. Third, it accomplishes its effect by portraying a world without trade-offs, as if niceness will
invariably benefit and selfishness will invariably harm both self
and others. For anyone who steps into a belief system such as
this, it is easy to decide what to do. Intelligence (or another
process of selection) might have gone into the creation of the
belief system, but not much intelligence is required to be
guided by it. Just follow the sign that says "This way."

Unsurprisingly, this "species of thought" is especially common in religious texts. The Hutterite text quoted in Chapter 28 is completely polarized when the words are classified
according to my categorization scheme. One list of words refers
to behaviors that are good for both self and others: brotherli-
ness, community, discipline, faithfulness, love, mutual help,
obedience, order, sacrifice, surrender, true equality. These behaviors produce a shower of benefits: food, shelter, "the most
precious jewel," "the hidden treasure," "true life," "enduring
life and joy," "genuine happiness," "hundred-fold profit," and
"power that gives life to bear fruit." There is no need to make
clear distinctions between material and psychological benefits because everything is thrown into the bargain. Another
list of words refers to behaviors that are bad for both self
and others: arrogance, avarice, covetous desires, ego, greed,
individuality, pride, selfishness, self-interest, self-seeking, self-
will. There are no words—not a single one—referring to
behaviors that benefit self at the expense of others or others
at the expense of self over the long term. The Hutterite belief
system portrays a world without trade-offs, a single track in
which the only rational decision for the believer is to sprint
toward glory and away from ruin. The Red Cross slogan "Do

something for nothing. And you'll get everything" is a perfect distillation of what the Hutterite text portrays in more elaborate terms.

It might seem amazing that the familiar definitions of selfishness (benefiting self at the expense of others) and altruism (benefiting others at the expense of self) are totally absent from this particular religious text. The distinctions that philosophers love to dwell upon are similarly absent. The idea that an "apparently" altruistic act is "really" selfish because it also benefits the actor materially or psychologically is foreign to the Hutterite imagination. Even more amazing, it is foreign to the imagination of all religions. This sweeping statement is based not on my own work but on an ambitious project funded by the John Templeton Foundation and carried out by three world-class religious scholars named Jacob Neusner, Bruce Chilton, and William Scott Green.

The Templeton Foundation is interested in the concept of altruism, along with its interest in forgiveness that funded my project resulting in *Darwin's Cathedral*. They approached Jack Neusner to organize a conference on altruism and religion. Jack charged Bill with the task of formulating a set of questions about altruism to see if it is a useful and appropriate concept for the academic study of religion. These questions were then sent to experts on the major religious traditions, whose answers were presented at the conference and published in a book titled *Altruism in World Religions*, edited by Jack and Bruce. The resounding answer was that when altruism is defined as "intentional action ultimately for the welfare of others that entails at least the possibility of either no benefit or a loss to the actor," it is foreign to the imagination of *all* of the major religious traditions. Hinduism, Buddhism, Judaism, Islam, and Christianity all portray worlds without trade-offs in which benefiting others results in a shower of material, psychological, and otherworldly benefits for oneself. The word "altruism" wasn't even coined until the 1830s (by Auguste Comte) and represents a "species of thought" that is evidently confined to secular life. *Altruism in World Religions* ends with the following summary by Bill:

Although contemporary altruism seems not to find a home in the world's major religions, the results of this collective project are far from entirely negative. The inappropriateness of contemporary altruism to these religions does not mean that they are bastions of selfishness. The contrary is the case. The discrete studies in this volume display in wonderful detail how benevolence and charity operate within the values and structures of the religions studied. Whether in the Talmud's story of Mr. Five-Sins, or Jesus' parables, or the tale of Shaykh Abû ᶜAbd Allâh, or the teachings of Pali Buddhist texts, or the lessons of Nhat Hahn, or the discipline of Nicherin, or the teachings of Krishna, or the ideal of friendship in Greco-Roman philosophy, or the "graduated altruism" of the *Mencius*—every tradition represented in this book exhibits the important extent to which an active concern for others stands within its system and represents fundamental categories of meaning. Some of these examples bear some resemblance to contemporary altruism, and some do not, but as several of the authors point out, only by a rigid secular calculus is benevolence less benevolent because the actor benefits. Without question, religions are major forces for other-directed human behavior. That such behavior operates within a transcendent or eternal framework does not diminish its impact or lessen its capacity to improve the human condition.

These distinguished religious scholars did not need an amateur like me to tell them about the practical collective benefits of religion. Neither were they aware of my own lighthearted ramble through the English language, capturing and classifying word usages like so many butterflies. After we admire religions for their practical realism, however, we must also appreciate how they depart from factual realism along the way. The real world is full of complicated trade-offs, conflicts of interest, and win-lose situations. In principle, a belief system could score high on factual realism by representing all of these complexities and also score high on practical realism by showing how to deal with them. In practice, such a belief system would probably require too much time, energy, and mental resources for most people and societies to operate. Religious belief systems are more user-friendly. They reduce the complexity *of* the real world to motivate a suite of behaviors that are adaptive *in* the real world. Ironically, the reason that trade-offs are absent from religious beliefs systems is *because* of a trade-off between maximizing factual and practical realism at the same time.

My search for uses of the word "selfish" eventually led me to two books with nearly identical titles: *The Art of Selfishness* by David Seabury, written in 1937, and *The Virtue of Selfishness* by Ayn Rand, written in 1963. These represented a different "species of thought," as we can see from the following passage from Seabury: "Years ago, I decided to go abroad in preparation for my vocation. My mother was sixty-two. Eight of her friends wrote reminding me she was well along in years and pleading with me not to leave until she died. She passed away at the age of ninety-three. The writers of those letters condemned me as *selfish* because I left as I did. My mother suffered to have her wishes disregarded, but told me a few weeks before she died that one of the best things I had ever done for her was to leave her when and as I did. Had I not gone, it is obvious I would have begun my training in the fifties. I would have carried in my heart a grievance more hurtful to our relation than my absence. I could not have been the financial and spiritual support my profession made possible."

This use of the word "selfish" refers to a behavior (being inconsiderate) that is good for Mr. Seabury and bad for his mother over the *short* term but proves good for them both over the *long* term. The effects are both psychological and material. Selfishness has been turned from a win-lose proposition to a win-win proposition, the very opposite of the *Reader's Digest* story.

The main message of both books is that the conventional virtues are overrated and that the real secret of happiness is to follow one's own ideals, no matter how strong the opposition. Good advice, perhaps, just as religious advice might be good, but I was most interested in how it was presented.

As I read through Seabury's book, I collected a long list of words and phrases portrayed as good for everyone: cooperation, higher selfishness, constructive selfishness, integrity, knowledge, men of science, mutual aid, natural decisions, true unselfishness. Others were portrayed as bad for everyone: apparent unselfishness, coercive goodness, cruel virtue, denial, duty, pseudo-morality, pseudo-unselfishness, quasi-unselfishness, responsibility, sacrifice, self-denial, serfs of virtue, smirking self-sacrifice, tradition, unselfishness, virtuous conventionalists. Not a single word or phrase referred to behaviors that benefit self at the expense of others or others at the expense of self over the long term. Seabury's world is as polarized as the Hutterite world, impelling anyone who takes it seriously to sprint toward glory and away from ruin. The two belief systems promote different behaviors, but in exactly the same way.

Any belief system that glorifies selfishness must distance itself from obviously antisocial behaviors such as rape and murder. These are portrayed as stupid, not selfish, because their long-term effects are bad for self in addition to others, just like the conventional virtues: "But let us understand each other. This new liberty is not anarchy. There is no counseling of greed, of lust, of licentiousness in the attitudes of science. . . . Riotous arrogance and rapacious rebellion are not constructive selfishness—they are insanity." How can we tell the difference between good and bad selfishness? The principles of science, of course!

> There are, I suppose, four sorts of men on earth; ruthless egotists who take the way of greed; virtuous conventionalists, who follow the creeds; the blind rebels, who will not yield to any rules; and the men of science, who strive to obey natural law. There is no meeting point between the old and the new attitudes in the face of life's problems. We go two roads. Those who revere the "good old ways" follow the precepts and the conventions. Those who seek to obey nature, through the discoveries of science, follow another set of values.

It would be hard to find a clearer example of science being used as a substitute for God. In case the message of the book is obscured by the subtlety of the prose, Seabury spells it out in the prologue: "Do this and you will discover, as I did, that what is good for you is invariably good for others."

The Art of Selfishness was written in 1937, reprinted in 1964, and is currently out of print. Touchingly, a few devoted readers have still posted comments on Amazon.com testifying to how it changed their lives: "A pearl of wisdom"; "The content is with me to this day"; "We feel our happiness is due in large part to the guidelines we learned from this book." Seabury wrote in the style of a self-help author with few intellectual pretensions. By contrast, Ayn Rand is regarded as an intellectual giant by her followers and is disparaged with equal zeal by her detractors. Her most widely read book is *Atlas Shrugged*, which has over 1,300 readers' comments on Amazon.com. How many other books written in 1957 can make this claim? *The Virtue of Selfishness* is a collection of essays published in 1963 to explain and develop her philosophy of objectivism. It is still in print, and the cover of the current paperback edition shows a lone individual standing godlike atop a golden staircase facing a rising sun.

Despite its philosophical ambitions, I quickly discovered that the structure of *The Virtue of Selfishness* is identical to the homespun wisdom of *The Art of Selfishness*. The conventional virtues are shockingly placed in the lose-lose column with a long list of words and phrases: altruism, collective, faith, incomprehensible duty, moral cannibalism, mysticism, sacrifice, self-denial, self-immolation, self-sacrifice, unselfishness. The pursuit of self-interest is placed in the win-win column with another list of words and phrases: cooperation, egoism, honesty, independence, integrity, logic, pride, rational principles, rational self-interest, reason, self-esteem, selfishness. Finally, a list of stupidly selfish behaviors joins the conventional virtues in the lose-lose column: animal, blind desires, feeling, hedonism, irrational emotions, looter, mindless brute, moocher, Nietschean egoists, parasites, subhuman creature, urges, and whims. Like Seabury, Rand does not want her message to get

lost in the subtlety of her prose, so she spells it out directly
that "there are no conflicts of interest among rational men."
For example, if two rational men are competing for the same
job, they will agree on who is best and the other will voluntarily withdraw. You might protest that this is not what other
people mean by the word "rational," but what else should you
expect? Consistency is only expected within and not across
belief systems, as we have already seen for the word "selfish."

These two books have the same basic structure, not because one person copied from the other but because they
must to motivate the same suite of behaviors. They are like
two biological species that have independently adapted to
the same environment. In fact, a few differences between the
two books provide evidence for their separate origins. Any
belief system that exalts selfishness must have two categories
of selfishness, but the actual words used to distinguish them
are somewhat arbitrary. Seabury used the word "egoism" for
the bad kind of selfishness, but Rand used it for the good
kind, subtitling her book "a new concept of egoism." This
difference would not exist if Rand had copied directly from
Seabury, unless she was on the lookout for linguistic detectives
such as me and deliberately covered her tracks!

So far I have provided a dispassionate outsider's view of
polarized belief systems. What is it like to be on the inside
looking out? Nathaniel Branden provides a revealing glimpse
in his book on the Ayn Rand movement titled *Judgment Day*.
Branden first read Rand's novel *The Fountainhead* at the age
of fourteen. His older sister was giggling over the sex scenes
with her friends, and when they left he picked it up out of curiosity. His life was changed forever. The prose wove a spell
over him. As he described it to her later, "My excitement wasn't
just at the stylization of the writing—your particular way of
seeing and re-creating reality, which runs through everything—
it was like being in a *stylized universe*." That is a fine description of how a belief system can become more powerful by
departing from factual realism.

An intense and introspective youth, Branden had visions of
greatness for himself that were out of place in his middle-
class conformist world. Rand's book spoke to his innermost

desires: "I was aware that Ayn Rand had reached me in some unique way, and that in the cardinal virtues of the novel—independence, integrity, love of one's work, and a sacred sense of mission about one's life—I had found a world more interesting, more energizing, more challenging, and in a way more real, than the world around me. I experienced it as more relevant to my growth and development than anything I was hearing from my elders. To me, at fourteen, the vision offered by *The Fountainhead* was a great and inspiring gift."

During the next four years, Branden read *The Fountainhead* almost continuously, "with the dedication and passion of a student of the Talmud." If you read him a sentence, he could recite the preceding and following sentences. He described it as a "shield" and "fortress" against the mediocrity of his actual life. When he attended the University of California at Los Angeles, he and a circle of friends who were also devotees became her disciples, convinced that they were going to transform the world: "This is how we were back then, Ayn and I and all of us—detached from the world—intoxicated by the sensation of flying through the sky in a vision of life that made ordinary existence unendurably dull."

In short, the Ayn Rand movement had all the intensity of a religious movement. It didn't matter that there were no gods or afterlife. Salvation by rational choice proved just as intoxicating. It didn't matter that the texts were fictional and no one was being asked to believe that the characters had existed or the events had taken place. They were *better* than real because of the way that they organized perception, providing a shining path toward glory, a golden staircase to face toward the rising sun. Rand's disciples could pursue their individual goals with a clear conscience because everyone else was going to benefit as well. If people disagreed, they were easy to dismiss as conventionalists who suffered from "bad premises." If people suffered, then those were short-term effects and even they would thank you in the end. If you stumbled, then all you had to do was rethink your premises and you'd be on your way again. One member of the early movement was Alan Greenspan, the economist who was later to head the United States Federal Reserve Board. As he put it: "What she did ... was to make me think why capitalism is not only efficient and practical, but also moral." There's that connection between morals and morale again, just like the effect of Calvinism on the good citizens of Geneva.

Unfortunately, the discussions that took place within the inner circle were nothing like reasoned dialogue. Rand was regarded as infallible, like the oracle at Delphi, and would amaze her followers by providing an answer to any question with utmost certainty. It was as if she possessed some vast formula for solving all of life's problems. Her motto was "Judge, and prepare to be judged," so they spent much of their time analyzing each other to cast out false premises, like so many demons. She was highly vindictive, so those who didn't agree with her were quickly excluded as "subhuman creatures." When she and Branden initiated a love affair, they rationalized it to themselves and their spouses as noble and right. Years later, when Rand tried to resume the affair and Branden had fallen in love with another member of the group closer to his age, she cast him into the abyss in a towering rage.

The saddest proof of Rand's detachment from reality came when she discovered that she had lung cancer. A lifelong smoker, she was nevertheless shocked and kept it a secret from all but her closest friends. How could *she* get lung cancer when she had no bad premises? She was no more rational about the nature of disease than a Christian Scientist or those who line up to be healed at a revival meeting.

During Rand's funeral, a six-foot-high floral dollar sign was erected by her coffin and guards were on hand to keep out enemies such as Nathaniel Branden. She is no longer with us, but you can still enter her world through her books and numerous organizations such as the Ayn Rand Institute (http://www.aynrand.org), which issues the following appealing invitation: "Those who have read *The Fountainhead* or *Atlas Shrugged* know that the sunlit universe Ayn Rand depicts in her novels is unlike the world that they see around them. How can one achieve the clarity of vision and joyous existence that her fictional heroes achieve?"

What makes some people hate religion? Let me count the

ways: the way that religious groups hoard their benefits for their own members, their intolerance of other groups and even their own members who step out of line, and their wanton tendency to sacrifice factual reality on the altar of their own narrow purposes. For anyone who shares these concerns, I'm sorry to report that the situation is much worse than you think. All of these problems exist for belief systems that we don't recognize as religious, as we can clearly see in the case of the Ayn Rand movement. If you worry about religions, then you should *really* worry about stealth religions. The presence or absence of supernatural agents—a particular distortion of factual reality—is truly just a detail.

Nevertheless, I'm not out to trash stealth religions, any more than I was with the religions that I discussed in the previous two chapters. A more sympathetic rendering is required to understand the true nature of the problem. In the first place, practical realism is a *good* thing. If a belief system doesn't enable us to thrive, it should be exchanged for another that does. My commitment to factual realism is based on a belief—faith, if you will—that the most practical belief system for a large-scale society in the long run is one that is firmly anchored in factual reality. First we must know how the world really works (such as the causes of cancer) and then we must make wise decisions on the basis of the information.

Factual realism is not benign by itself, but only when joined with a benign value system. Facts by themselves can be used for good or ill. Values by themselves can't achieve their goals without facts. Put them together, and you have a belief system in which factual realism contributes to practical realism. The belief system will be benign at a large scale and in the long run only if it tallies the costs and benefits for many people and far into the future. Any value system that is restricted to a subset of people and concentrates on short-term benefits will ultimately wreak havoc, even for the very people who thought that they were benefiting themselves, like those cancer cells described in Chapter 19, who were doing just fine until they brought the life of the organism to an end.

This might be the ideal belief system to work toward in

the future, but it is absurd to hold the belief systems of the past and present to such a lofty standard. If the ideal belief system is possible at all, it requires large differentiated societies with the wealth and stability to pursue long-term goals rather than living for the moment. Establishing facts is a straightforward activity, as I have stressed throughout this book, but it is also hard work and an unaffordable luxury if you are looking for your next meal or fighting for your life. Throughout history, people have been faced with the problem of surviving in small groups over the short term with limited resources. The ideal belief system would never survive in such an environment. The belief systems that do survive are not *entirely* detached from factual reality. According to anthropologists, all cultures include a mode of practical reasoning that we easily recognize as pragmatic, rational, and protoscientific. Nevertheless, this mode is easily eclipsed by other modes that freely distort and make up facts to motivate successful behaviors. As evolutionists, this is exactly how we should expect the human mind to work.

Indeed, when you're an evolutionist it becomes obvious that deception begins with perception. All creatures perceive only those aspects of the physical world that are important for their survival and reproduction. Everything else becomes literally invisible. We can't navigate by the earth's magnetic field like some birds or detect prey by their electrical fields like some fish. Bats and elephants make sounds that we can't hear. Birds, insects, and flowers paint themselves brightly in colors that we can't see. Our ancestors departed from factual realism in their drive toward practical realism long before they evolved the capacity to have beliefs.

Those who criticize religions, and who need to worry even more about stealth religions, are right about their failures. However, they seem to think that just by pointing out the failures, right-minded people will see the light and the problem will be solved. That's just plain dumb. They need to understand that the problem is deeply embedded in the way we are as a species, and the solution requires creating a social environment in which their ideal belief system, and mine, can survive.

The Social Intelligence of Nations, or,
Evil Aliens Need Not Apply

IMAGINE STARTING A CONVERSATION with someone
who turns out to be nakedly selfish. Not only does she care
only about herself, but she *talks* to you as if she cares only
about herself. You hear all about her schemes for getting
ahead while riding roughshod over other people. Then she reveals how *you* figure in her plans without the slightest concern for your welfare. You would be doubly amazed at such a
person. Only a social idiot would first have such shrunken
values and then make no attempt to disguise them.

Yet I have just described the average political speech on international relations. It is amazing how politicians can stand
in front of their constituents and say without blushing that
the one and only thing on their mind is the welfare of their
nation. I'm not worried about the possibility that they might
be lying to their constituents. I'm worried about the possibility that they might be telling the truth. Don't they know that
members of other nations might be listening in on the conversation? We would be shocked at such a low degree of social intelligence in an individual, yet it is standard fare for a
nation.

If I were to tell the average politician that evolutionary
theory has something to say about international relations, he
would look at me as if I were from outer space. Yet nations are

nothing more than very large groups that are trying to function as collective units. The last fourteen chapters have been
all about groups evolving to function as collective units. Even
the individuals that our disbelieving politician counts among
his constituents are groups of groups of groups. The only
reason that we call them individuals is because they have
evolved to function as collective units so well.

Nations not only are composed of individuals but also include subgroups of all manner and description. The smallest
are the size of hunter-gatherer groups and miraculously constitute themselves, as Tocqueville noted so perceptively, not
because it is easy but because it is automatic, like opening our
eyes and seeing. These were the *only* human groups a mere
fifteen thousand years ago (give or take a few thousand).
Cultural processes had already shifted human evolution into
hyperdrive, enabling us to populate the globe and occupy
hundreds of ecological niches, but group size was limited primarily by the availability of resources. Once that limiting factor was removed by the invention of agriculture, the scale
of human society began to increase by the same process of
multilevel selection that gave us bodies, beehives, and small-
scale human societies. The larger groups managed to function
as collective units by supplementing our innate "social physiology" with new mechanisms that had never existed before.
Even the city of Geneva in the 1500s, which was minuscule
compared with a modern nation, required the invention of
new mechanisms, as I described briefly in Chapter 28 and at
more length in *Darwin's Cathedral*. Some of these mechanisms were consciously invented but others just emerged
from the raw process of cultural evolution, as the few inadvertent social experiments that managed to hang together,
compared to the many that fell apart.

In short, modern nations represent the current frontier of
multilevel cultural evolution. The ideas in this book can provide a powerful framework for understanding how a group
the size of a nation can function as a collective unit and how
interactions among nations can be guided in the direction of
cooperation rather than conflict. Before proceeding, however, a word of caution is in order. Political theorists are only

beginning to encounter the ideas that have become familiar to you. A book on political cultural evolution comparable to *Darwin's Cathedral* has yet to be written. Nevertheless, some observations are so basic that I would be surprised if they prove to be false. Moreover, it is important to show that evolutionary theory can be used to advance political ideals that are the opposite of Social Darwinism as imagined in the 19th century. With these caveats in mind, I offer the following nuggets of homespun evolutionary wisdom about the biggest and dumbest organisms to ever walk the face of the earth.

We are not fated by our genes to engage in violent conflict. Long before Darwin, our bloody past has been cited as evidence for the inevitability of a bloody future. Even though this argument is often stated in terms of genes, it makes no sense from an evolutionary perspective because no single behavior is adaptive across all environmental conditions. Bloody conflict is not everywhere. It has a "distribution and abundance," like a species in an ecosystem. It does indeed thrive under some conditions but loses to more benign strategies under other conditions. The Vikings of Iceland were among the fiercest people on earth, and now they are among the most peaceful. In principle, it is possible to completely eliminate violent conflict by eliminating its preferred "habitat," regardless of how rare or common it has been in the past.

The average person is a facultative sociopath. We might not be *fated* by our genes to engage in violent conflict, but we are most certainly *prepared* by them. I do not wish for a moment to downplay the role of violent conflict in human genetic and cultural evolution. In addition to a rich archeological, anthropological, and historical record of murder and mayhem, there is ample psychological evidence that we are hardwired to distinguish between "us" and "them" and to behave inhumanely toward "them" at the slightest provocation, as science journalist David Berreby recounts in his book *Us and Them: Understanding Your Tribal Mind*. Much of this research was initiated in the aftermath of World War II to explain how decent people could have participated in the Holocaust. Henri

Tajfel, a Holocaust survivor, discovered that he could trigger us-versus-them thinking merely by assigning people to arbitrary groups. In what has become known as the Robbers' Cave experiment, social psychologist Muzafer Sherif and his colleagues showed that well-bred American boys at a summer camp could be easily set against each other by housing them in separate cabins and reunited just as easily by giving them a common task. Contemporary books such as *Among the Thugs* by Bill Buford and *War Is a Force That Gives Us Meaning* by Chris Hedges describe violent conflict as instinctively pleasurable, like a sexual experience. In earlier chapters I described some adaptations as like elaborate "war plans," ready to be implemented at a moment's notice. The war plan metaphor includes not only metabolic strategies triggered early in development, as described in Chapter 8, but real war plans, triggered all too easily whenever they appear warranted. My dictionary defines a sociopath as "a person with a personality disorder manifesting itself in extreme antisocial attitudes and behavior." An ounce of evolutionary thinking makes it obvious that "extreme antisocial attitudes and behavior" do not necessarily reflect a personality trait or a disorder, but may be an evolved tendency that can be expressed by anyone in the grip of us-versus-them thinking. If we want to avoid this kind of facultative sociopathy, we need to avoid pushing the wrong psychological buttons, just as the president of the United States avoids pushing the fabled red button that is supposed to initiate a nuclear strike.

Watch out for the invisible hand. In *The Theory of Moral Sentiments* (1759) Adam Smith famously claimed that individuals who care only about their own narrow interests are led by "an invisible hand" to benefit society. Bernard Mandeville expressed the same idea in his *Fable of the Bees* (1705), which humorously portrays human society as a teeming beehive whose members act only for personal gain. Two concepts must be distinguished to evaluate these metaphors in modern evolutionary terms: self-*organization* and self-*interest*.

A system is self-organized when its component parts interact in a relatively simple manner to produce complex and un-expected ("emergent") behaviors at the level of the whole system. Clouds, tornadoes, and snowflakes are examples of self-organized physical systems. All life forms are self-organized. A real beehive provides a fabulous example of self-organization at the group level, as I described briefly in Chapter 20 and Tom Seeley describes in much more detail in his book *The Wisdom of the Hive*. Each bee follows a relatively simple set of behavioral rules such as "pick a dancer at random" or "dance less if you had to wait a long time to unload your nectar," but these rules combine to produce the complex social physiology of the hive. If that is what Smith and Mandeville were driving at with their metaphors, then they were definitely on the right track for human groups. Our social physiology is like an iceberg with only a tiny fraction above the surface of conscious awareness, as I have already stressed in previous chapters.

However, Smith and Mandeville also claimed that the conscious intentions of people could be described as "self-interested." Here is how Mandeville put it humorously to verse in his fable:

> As Sharpers, Parasites, Pimps and Players,
> Pick-pockets, Coiners, Quacks, Sooth-Sayers,
> And all those, that, in Enmity
> With down-right working, cunningly
> Convert to their own Use the Labour
> Of their good-natur'd heedless Neighbour:
> These were called Knaves; but, bar the Name,
> The grave Industrious were the Same.
> All Trades and Places knew some Cheat,
> No Calling was without Deceipt.

Smith's prose is more ponderous but makes the same point:

> The rich only select from the heap what is most pre-cious and agreeable. They consume little more than the poor, and in spite of their natural selfishness and rapac-ity, though they mean only their own conveniency,

though the sole end which they propose from the labours of all the thousands whom they employ, be the gratification of their own vain and insatiable desires, they divide with the poor the produce of all their improvements. They are led by an invisible hand to make nearly the same distribution of the necessaries of life, which would have been made, had the earth been divided into equal portions among all its inhabitants, and thus without intending it, without knowing it, advance the interest of the society, and afford means to the multiplication of the species. When Providence divided the earth among a few lordly masters, it neither forgot nor abandoned those who seemed to have been left out in the partition. These last too enjoy their share of all that it produces. In what constitutes the real happiness of human life, they are in no respect inferior to those who would seem so much above them.

This particular passage does not fully represent Smith's views about morality, since he disagrees with Mandeville and emphasizes more other-oriented instincts elsewhere in his book. In any case, the modern interpretation of both metaphors is that people are innately self-interested, that the concept of self-interest can be reduced to something like the utility maximization of economic theory, and that self-interest robustly leads to well-functioning societies. This interpretation is deeply flawed on the basis of elementary evolutionary principles that are very unlikely to be wrong. I have written an article for the experts about this titled "The New Fable of the Bees," which is available from my Web site, but the gist of my argument can be easily understood on the basis of this book. There are billions of ways for a human society to be self-organized. Of these, only a tiny fraction are adaptive in any sense. Of these, only a fraction are adaptive at the group level. The simple rules that enable people to organize into adaptive societies evolved by cultural and genetic group selection over a long period of time. Other simple rules that tend to disrupt society evolved by within-group selection. The rules that exist in one society need not exist in another,

as I stressed for religions in Chapter 29. There is no way to reduce all this complexity to a simple minimalistic formula
such as self-interest or utility maximization. Mandeville and
Smith were right about self-organization but wrong about
self-interest. We will never understand the relatively simple
rules that cause people to self-organize into adaptive societies until we study them from an evolutionary perspective.

Evil aliens need not apply, part 1. How many times have
you heard the conjecture that worldwide cooperation can be
achieved only in the face of an invasion by aliens from outer
space? Not only is this suggestion impractical, but it probably
wouldn't work. Remember that evolution is all about *differences* in survival and reproduction, regardless of long-term
consequences. The cancer cells described in Chapter 19
spread at the expense of their solid citizen neighbors, despite
hastening their own destruction. If the earth was faced with
global destruction (and it *is*—it's just harder to visualize than
scary aliens), some of us would surely sacrifice more than
others, some would engage in rampant profiteering, some
would help the aliens in an effort to save their own skins, and
so on. Increasing the scale and intensity of the threat would
not eliminate these differentials.

Evil aliens need not apply, part 2. Another version of the evil
alien argument is to portray one's adversaries on earth as evil,
such as the Soviet Union as an "evil empire" or Iraq, Iran, and
North Korea as an "axis of evil." These two examples happen
to involve American presidents belonging to the Republican
party, but I am not singling out any administration, party, or
nation. This kind of language is designed to bring out the facultative sociopath in all of us by dehumanizing our adversaries. Our political leaders are cautious about pushing the
fabled red button that is supposed to initiate a nuclear strike,
but they bang away at our psychological buttons all the time.
In personal interactions, we avoid people who act impulsively
and cannot keep their emotions under control. We value people who remain cool, calm, and collected in tense situations.
A nation should be held to the same standards.

Short-term emotional responses cannot be sustained and become toxic over the long term. Our emotions are designed to mobilize our bodies and minds for short-term events and cannot be sustained over the long term. Disasters frequently result in a spontaneous outpouring of help, accompanied by deeply pleasurable feelings of unity, but then people return to their baseline emotional state within a matter of weeks. The emotions of fear, anger, and hatred have such powerful effects on our bodies and minds that they are literally toxic over the long term, eating away at our immune systems and even our brains, as my evolutionist colleague Robert Sapolsky recounts in his book *Why Zebras Don't Get Ulcers.* A national policy or any other belief system that attempts to sustain strong emotions such as fear and hatred over the long term is almost certain to fail and to produce severe negative side effects over the long term. Human potential can be developed only when we are not scared, angry, or hungry. Our evolution as a species required periods of safety and satiety, which we recognize and communicate through laughter, as we saw in Chapter 23. If we aren't laughing and enjoying each other's company, we aren't developing our potential.

Taking the global village seriously. If the invisible hand and evil aliens can't make us get along, what can? Marshall McLuhan famously observed in 1964 that rapid travel and electronic communication are turning the world into a global village. This metaphor deserves to be taken very seriously. A real human village is just the right size for our social instincts to work effectively, as I have stressed again and again. If we can scale up village interactions, with the nations as the individuals and the planet as the village, worldwide cooperation will become so natural that it constitutes itself, to use Tocqueville's felicitous phrase. However, this will require more than rapid travel and electronic communication. In addition, we need to provide the conditions that initially caused human society to cross the cooperation divide, as outlined in more detail below.

The global village cannot consist of village idiots. Forming a treaty with a nation, holding a nation accountable for an act of terrorism, or any number of other political decisions assumes that the nation can be treated as a "corporate unit" and held accountable as if it is an individual. The more a nation is torn by internal conflicts, the more idiotic it becomes as a corporate unit. Chinua Achebe's novel *No Longer at Ease*, which I used to illustrate the difference between oral and literate thought in Chapter 27, also reveals the nature of corruption in a nation such as Nigeria. The traditional corporate unit is a single village, such as Obi's village of Umuofia. The tribe consists of a collection of villages that regard each other as competitors, roughly like the teams of a sports league. No one in Umuofia regards himself or herself as Nigerian, and the inhabitants are proud to be the first of the villages to send one of their sons overseas to be educated. They regard placing one of their sons in a position of power in the capital city of Lagos as the new field of competition that has replaced raids and feasts. *Of course* such a person should act as a patron for his village. A fundamental tenet of multilevel evolution is that virtues at one level become vices at higher levels; looking out for oneself becomes selfishness, looking out for one's family becomes nepotism, and looking out for one's village becomes corruption. People do not change their social identity by decree. A long historical process was required for the inhabitants of Europe to regard themselves as British, French, and Germans rather than members of smaller social units. An individual such as Obi can make the transition, proudly regarding himself as Nigerian and vowing to resist corruption, but it is vastly more difficult to change a geographical region whose various subpopulations are so isolated that they can't speak each other's language. When Obi succumbs to corruption, he has yielded to the moral imperatives of society at a smaller scale. Before a country such as this can act as a responsible member of a global village, it must solve these internal problems so that it can begin acting as a corporate unit.

Where there is no discipline and excommunication, there is no community. This statement comes from the Protestant reformer Martin Bucer quoted in Chapter 28, with the word "Christian" removed. Once the members of the global village are capable of acting as corporate units, there must be a relatively equal balance of power. This makes all the difference between a despotic chimplike society and an egalitarian small-scale human society, as we saw in Chapter 21. There must be a way to control upstarts and bullies, for nations no less than individuals. Anthropologist Steve Lansing's book *Priests and Programmers* is mostly about the Water Temple religion of Bali, as I briefly described in Chapter 28, but it begins by recounting the brutal exploitation of the Balinese by the Dutch. The history of colonialism provides ample evidence that there is no such thing as an enlightened nation when it comes to between-group interactions. Egalitarianism must be *guarded*, for the global village no less than a real village.

The most powerful nations should learn the virtue of humility. The balance of power is never exactly equal in real villages. Some members are always better than others in their physical prowess, intelligence, or experience. A powerful member of a village signals his or her good intentions by adopting an attitude of humility, like the !Kung hunter described in Chapter 21 who humorously downplayed his prowess and allowed himself to be kidded by his compatriots. Powerful members of villages who demonstrate their good intentions are rewarded with leadership, while those who throw their weight around are shunned and excluded.

Morality is required for morale. According to Chris Boehm, small-scale societies are above all *moral communities*. They are unified not by having the same genes but rather by having the same sense of what is right and the means to enforce it. A strong value system is required for decisive action. This theme has arisen repeatedly, from Calvin's Geneva to Alan Greenspan's free-market economy. It also makes sense from a purely theoretical perspective. In Chapter 2, I said that facts never lead to actions all by themselves. They can only inform a value system. If there is no agreement on the value system, then there is no agreement on what to do on the basis of facts. A shared value system is therefore required at the international scale for the global village to become a moral community.

Moral communities can embody whatever values are deemed good and right. Moral communities have at least two dark sides. They do not necessarily extend their blessings outside the moral circle, and the way that they enforce their norms within the moral circle can be repressive and even brutal. If you have access to the Internet, type "stoning" into Google Images and you will see what I mean. On a milder scale, I know plenty of people who hated the pressures to conform in their small towns and felt liberated when they moved to a place such as Los Angeles. These aspects of morality are themselves immoral for some people. However, the enforcement of norms need not be repressive and need not stifle forms of diversity that are valued by members of the moral community. Pluralism can be enshrined as a virtue and its suppression punished, for example. Most of the small-scale societies reviewed by Chris Boehm value autonomy as much as conformance to norms that are agreed upon by consensus. One of the strongest prohibitions is against meddling in other people's affairs. Virtually any value system can be stabilized by rewards and punishments, as long as it is agreed upon by consensus.

Learn from the wisdom of religions and other traditional social organizations. Science is accorded such respect in modern life that those who claim scientific authority are often as arrogant as the leaders of powerful nations. They assume that current scientific knowledge is superior to knowledge derived from any other process. The scientists who claimed that higher education interferes with ovarian development were just as arrogant. Science has been dead wrong in the past, is dead wrong about some things now, and will be dead wrong about other things in the future. Scientific methods enable scientists to discover when they are wrong and to labor toward the truth, with many twists and turns along the way. As for knowledge that is regarded as nonscientific, when it is the product of evolutionary processes, it can vastly surpass our current scientific fumbling. The concept of society as an organism provides an excellent case in point. It was regarded as common sense by scientists during the early twentieth century, then it was abandoned, and now it has been newly rediscovered. So far science has been like a blood-hound having difficulty finding the scent. In the meantime, there are real human societies that have been functioning as organisms for centuries and millennia. The appropriate stance for scientists to adopt is that traditional societies are likely to embody a great deal of wisdom that remains to be discovered scientifically. They need to probe traditional societies for their wisdom in the same way that they probe Dicty the cellular slime mold and other model organisms in biology. That is how Sir John Templeton decided to spend his millions, and other funding agencies would do well to follow suit.

The best way to design a large-scale society is to understand in detail how societies work at a smaller scale. I am impressed by how the Hutterites respond to deviant behavior, by starting with gentle admonishments and escalating only as necessary. I am especially impressed by how their social system is designed to punish *behaviors* rather than *individuals*. In modern American society, a person who behaves badly is often assumed to be a bad person without any capacity for change. This belief is dressed up in scientific clothes when we look for the "genes" for behaviors as if there should be a one-to-one correspondence. In Hutterite society, individuals who behave badly are assumed to be capable of change and are welcomed back into the society as soon as they demonstrate convincingly that they have changed. I am equally impressed by how a concept such as the sacred can be understood in purely utilitarian terms, as I described for my own family in Chapter 29. And how about the power of dance? Could we establish world peace if everyone at the United Nations showed up in leotards?

Do not attempt social transplant surgery unless you are qualified and have the consent of the patient. If societies are like organisms with complex physiologies, then fundamentally changing a society is like performing a major organ transplant. It requires detailed knowledge and certainly the

consent of the patient. Powerful nations such as Britain during its colonial period and present-day America feel certain
about their vision of society and how to implement it in
other nations. In reality, they are like ignorant doctors who
think that they can strap a patient to the operating table,
yank out the existing organs, slap in some new organs, and
send the patient out the door expressing gratitude. Guess
what? It doesn't work. I have already quoted the character of
Obi in Chinua Achebe's novel *No Longer at Ease*, who rumi-
nates this way about the English: "Let them come and see
men and women and children who knew how to live, whose
joy of life had not yet been killed by those who claimed to
teach other nations how to live." Another character is a British colonial administrator named Mr. Green, who arrogantly
makes judgmental statements about Nigerians throughout
the novel, which ends with the following words: "Everybody
wondered why. The learned judge, as we have seen, could not
comprehend how an educated young man and so on and so
forth. The British Council man, even the men of Umuofia,
did not know. And we must presume that, in spite of his
certitude, Mr. Green did not know either."

You might agree with some of these observations and disagree with others. Some might strike you as self-evident, others as perversely wrong. They don't fall cleanly into any
current political camp; some sound liberal and others conservative. My claim is that all of them follow from evolutionary
theory at such a basic level that they are unlikely to be wrong.
Moreover, they are a far cry from the justification of social inequality that became associated with social Darwinism in the
nineteenth century. I could be badly mistaken and deceived
by my own biases. All factual claims, including my own, need
to be held accountable by scientific methods. I look forward
to the day when evolutionary theory becomes part of the basic training for all people who study and run our governments
and economies. Perhaps it will help us increase our collective
social intelligence so that we can manage our affairs more
successfully in the future than in the past.

32

Mr. Beeper

THE LAST FIFTEEN CHAPTERS have taken us on an extended journey, from the origin of life and the social behavior of microbes to the nature of religion and the fate of nations. This kind of integration is what Darwin meant when he referred to his theory as "one long argument." I enjoy thinking about the big questions from an evolutionary perspective, the kind that philosophers like to ponder, such as the nature of religion or our ability to comprehend reality. There is also something unsatisfying about such lofty questions, however, like taking a trip into space and wanting to return to earth. I look back fondly upon my earlier days when I studied more down-to-earth questions, such as how burying beetles regulate the size of their brood, or how highly sensitive sunfish manage to survive and reproduce. There is nothing like kneeling on the forest floor to excavate a burying beetle chamber, or floating on the surface of a pond with a mask and snorkel to observe the denizens below. In addition to the sheer pleasure of being outdoors, there is the pleasure of being certain about something small, if humble, rather than uncertain about something large, if grand. Anne, who still does this kind of work, has never quite forgiven me for leaving it to join the philosopher's table. Where is the man that she married?

When we moved from our farm in Michigan to Binghamton

University, we decided to live in town but still wanted a piece of land in the country. Fortunately, real estate prices in our vicinity enabled us to afford both on our professors' salaries. Our land is twenty miles from our house and enables me to ramble around forests, fields, swamps, a pond, and a beautiful stream, without studying anything in particular. After we had been here for a while, an article appeared in *Smithsonian* magazine on some of the wonderful tree houses that have been built around the world. One of Anne's graduate students, Michelle Berger, had a partner named Kevin Bach (now her husband) who is a master builder. Kevin had always wanted to build a tree house, so he showed us the article, and we couldn't resist. That is why I am sitting 20 feet up in a tree right now, or rather in a cabin built within a triangle of trees, complete with a woodstove, a sleeping loft, and two balconies. It is January, so the stream is clearly visible to my right, unobscured by vegetation. To my left is a swamp where coyotes hunt wild turkeys. The place is surprisingly wild, even though it adjoins a road, because hardly anyone walks back here except during hunting season. The animals always appear genuinely surprised to see us, as if they are unaccustomed to being disturbed.

This is my favorite place to write and think. I wish I could say that I wrote my whole book up here, but the fact is that I can and do write anywhere, a skill that I picked up as a parent. This book was written over the course of a year at home, in my office, in airports, airplanes and hotel rooms, in bars and cafés, in half a dozen states and three continents. Have laptop, will travel. But whenever possible I come out here to write, interspersed with walks. One of my favorite animals that I encounter on my walks is the wood turtle (*Clemmys insculpta*), which lives near streams but spends most of its time on land. Some of the wood turtles here are in their fifties, just like me, and turtles were roaming the earth before the age of dinosaurs. I am strongly tempted to study wood turtles to learn more about life in the slow lane. I could easily outfit them with radio transmitters to find them whenever I want. They would become accustomed to my presence, as they have to other unthreatening large animals such as deer. I could come with my laptop and a lawn chair, recording their behavior at intervals as I do my other work. We could grow old together. I've read a bit about wood turtles and I'm sure that they have stories to tell. One report claims that they stomp on the ground in a way that entices earthworms to come to the surface to be gobbled up.

A few years ago I discovered an opportunity to study people in the same down-to-earth way that I used to study sunfish and would like to study wood turtles. I was invited to a workshop sponsored by the Templeton Foundation on the lofty philosophical concept of purpose. Can evolution be said to have a direction over the long term, other than adapting organisms to their immediate environments? The location was a resort in the Bahamas where Sir John had one of his residences. The participants would be few in number and Sir John himself planned to take part. How could I resist?

Whenever I travel to a place such as the Bahamas, I bring my mask, snorkel, and fins for the sheer fun of it. I therefore arrived with a duffel bag on one shoulder and my briefcase on the other to one of the wealthiest resorts that I have ever seen or heard about. I think it has something to do with the Bahamas being a tax shelter. In addition to the resort with guest rooms on manicured grounds by the beach, there were enormous mansions built all around, looking out of place and too close together, as if some alien spaceship had taken them, one from Tudor England and another from the antebellum South, to be warehoused on this sandy spit. Extremely rich people lived in these mansions and used the facilities of the resort. Most of them were extremely rich elderly people. I guessed that their average age was about 65 as I glanced about the lobby while checking in. The staff was entirely black and the clientele entirely white. I'm not a firebrand about social issues, but I couldn't help feeling guilty as the maid turned down my bed covers and oh so softly closed the door, leaving me in my palatial room with a balcony facing the sea.

The idea that evolution is taking us toward a final destination has a long history and comes in many varieties, but nowadays it is most closely associated with Pierre Teilhard de

Chardin, one of the most colorful figures in the history of evolutionary thought. Teilhard was a Jesuit priest and paleontologist, a combination that might sound strange until we remember that science and religion were much more closely affiliated in the past than they are now. He joined a number of fossil-hunting expeditions, including the one that discovered the early hominid specimen known as Peking man in 1929. He felt and thought deeply about the human condition, perhaps influenced by his experience as a stretcher bearer in World War I, witnessing the killing and maiming of thousands of men. He invented a belief system that is difficult to categorize in terms of science and religion. He thoroughly accepted evolution and rejected aspects of Christian doctrine that he regarded as superficial, but felt that evolution fulfilled Christian doctrine in a deeper and more abstract sense. The final sentence of his best-known work, *The Phenomenon of Man*, is: "In one manner or the other, it still remains true that, even in the view of a mere biologist, the human epic resembles nothing so much as a way of the Cross." Unfortunately, the Catholic Church didn't see it that way and censured Teilhard's radical views. *The Phenomenon of Man* was not published until after his death in 1955.

Teilhard thought that human consciousness represents a new level of life that he called the noosphere. In the past, humans were a passive product of evolution along with all other creatures, but now they could take control of evolution: "Man is not the center of the universe as once we thought in our simplicity, but something much more wonderful—the arrow pointing the way to the final unification of the world. This is nothing else than the fundamental vision and I shall leave it at that." The final unification, the target that the arrow was hurtling toward, was called the Omega point and was a kind of diffuse global consciousness: "We are faced with a harmonized collectivity of consciousnesses to a sort of super-consciousness. The earth not only becoming covered by myriads of grains of thought, but becoming enclosed in a single thinking envelope, a single unanimous reflection."

Whatever we might decide about the scientific merits of Teilhard's ideas, they have the inspirational quality of a religion, making the reader feel like part of a cosmic plan with an exalted future. The Omega point is a bit like the detached state of nirvana in Buddhist thought, but Teilhard's vision has the additional appeal of being scientifically grounded, or so he claimed. No wonder that *The Phenomenon of Man* is still read and discussed, much like the novels of Ayn Rand. Both are designed to shoot their readers like arrows from their inspirational bows, although in very different directions. No wonder the Templeton Foundation was eager to explore the merits of Teilhard's vision from a current scientific perspective.

The group that had been assembled to discuss this loftiest of subjects was stellar. It included Francis Fukuyama, the political theorist whose book *The End of History and the Last Man* had an Omega point ring to it; Martin Seligman, father of the positive psychology movement; and Robert Wright, the science journalist whose new book *Nonzero* was on this very topic and who had helped to organize the conference. Take a subject like this, add a dozen intellectual top guns, and it is like igniting jet fuel. Within moments after introducing ourselves to each other at the first meeting, we were airborne, weightless, and giddy. Is humanlike consciousness a lucky coincidence or an inevitable outcome of evolution? What does it mean for a group to have a collective consciousness? The only person who remained silent was Sir John Templeton, then in his eighties but still very alert, who listened without displaying much emotion. If I were to guess, he was thinking: "Now I have actually seen angels dancing on the head of a pin." Sir John is above all a practical man, as I knew from reading his own books and collections of wise proverbs from religions around the world. He didn't make his millions by speculating on the Omega point. Finally he lost patience and broke into the conversation with his twangy Tennessee accent.

"What I want to know is—how can we inspire our *young* people?"

The whole room fell silent at this down-to-earth question as we looked at him, not knowing what to say. Then we roared off again to continue exploring the cosmos.

As soon as there was a break in the intellectual action, I ran to my room, pulled on my swim trunks, and hit the beach. It was empty, despite the lovely afternoon. The elderly patrons of the resort had evidently given up sunbathing and water sports. There were no young people to inspire, other than one bored lifeguard staring at the horizon. No matter. I had a date with the fishes. How I love snorkeling in shallow water, speeding along with my fins, knifing down to the bottom for a closer look, and returning to the surface with a mighty blast of air and water through the snorkel! The animal watching was actually a disappointment. The water was turbid and shifting sand does not afford much of a purchase for life. By swimming along the shore I found a rocky shelf, however, teeming with five or six species of small fish and many other species of even smaller creatures, bustling about with their lives as if I didn't exist. I hovered above them, rocked by the undulating waves, until it was time to return for dinner. They surely had stories to tell and if I knew their names I could read about them from the articles of my evolutionist colleagues who still did this sort of down-to-earth research for a living.

One of my incentives for attending this workshop was to meet another participant named Mihaly Csikszentmihalyi. To his linguistically challenged American friends, he says, "Call me Mike." I knew a little about his work and contrived to sit next to him at dinner to learn more. Mike is best known to the general public for his book *Flow*, which is about the psychological experience of being totally immersed in what you are doing, such as an athlete who is "in the zone" or me when I am writing this book. The way that he studies flow and other subjects is by outfitting people with devices that beep at random times during the day. In the old days they were pagers of the sort that doctors used to wear, but nowadays they are preprogrammed electronic watches. People who participate in the research carry a booklet with them, which they fill out every time they are beeped. Each page of the booklet asks what time they were beeped, what time it is now (people cannot always respond immediately), where they were, what they were doing, who they were with, and what they were thinking. Then they fill out a checklist of emotional and mental states on a numerical scale. Were they enjoying what they were doing? Were they concentrating? Did they feel in control? Did they feel happy, active, cheerful, lonely, worried, angry, responsible, irritated, proud, friendly, and so on? Were they in pain? The questions require about two minutes to answer and provide a snapshot of the person's external and internal experience. The day is divided into two-hour blocks and they are beeped within each block once at a time that is chosen at random, so there is a relatively even coverage of the day but no way to predict exactly when a beep will occur. People participate for a week in most experiments, providing over fifty snapshots of their personal experience. This was how Mike studied the who, how, why, where, and when of peak psychological experience.

Mike is an affable man with a bearlike appearance who still speaks with a thick Hungarian accent. I liked him immediately and forgot all about my dinner as he told me more about his beeper method. He called it the "experience sampling method," but it was exactly what I and my evolutionist colleagues call "point sampling" in animal behavior research. If I were to actually study wood turtles on my property, I would check them at random times to make sure that my sample of their behavior is representative, just as I was so careful to obtain a random sample of religions in Chapter 28.

Mike's beeper method had additional features that made perfect sense from an evolutionary perspective. It examined people from all walks of life, as opposed to the college students that act as the guinea pigs for so much psychological research. It examined them in the "natural environment" of their everyday lives, as opposed to artificial laboratory experiments. It sampled their experience at the moment that it was happening, as opposed to relying upon their memory. Who among us can remember what we were doing, thinking, and feeling even yesterday, much less a week or a month ago?

Then there was the sheer scope of the studies that Mike had performed over the years. Literally thousands of people had participated, so many that even statistically rare events were represented such as heart attacks, falling out of windows, and committing suicide. That's right—at least some people actually filled out their booklets until the end. Furthermore, much of the data from past studies was stored on computer and could be made available to people such as myself with appropriate training. Mike had turned the beeper method into a veritable institute with a staff for maintaining the mountain of information that had been generated.

For me, Mike's beeper method was vastly more exciting than the lofty subject of purpose that we had been brought here to discuss, precisely because it was so down-to-earth. When the intellectual air show resumed the next morning, all I could think about was how I could use the information from Mike's past experiments to ask my own questions about human behavior from an evolutionary perspective. When I returned to the water to hover above the little fish on their rock shelf, I imagined myself hovering above the people who had taken part in Mike's experiments, bustling about their lives without any concern for my presence.

The end of the conference coincided with a gala banquet that the resort periodically staged for its guests and the occupants of the surrounding mansions. The guilty feeling that I experienced upon my arrival intensified as I entered the dining room with an orchestra of black musicians and an army of black servants tending a crowd of elderly white people. I took a plate, walked toward the food area, and stopped in my tracks. Never before had I seen such a lavish display of seafood, artistically arranged to please the eye in addition to the palate. There were colorful pyramids of shrimp and crabs in their shells. Fish were arranged in parallel as if they were still swimming in schools. I would have been thrilled to meet any one of them while snorkeling off the beach. I'm sure that they would have had stories to tell. Seeing them, lifeless, in such abundance and arranged so tastefully, made me think of the Chinese emperors who reputedly feasted on dishes made of hummingbird tongues. This was an Omega point that Pierre Teilhard de Chardin had not anticipated. I lost my appetite and returned to my room.

After the workshop I read everything that I could find about the beeper method. Mike and his colleagues had studied a variety of subjects in addition to flow, including the effects of watching television, what happens to children who are identified as gifted in school, and what happens to a marital relationship when the wife has a career of her own. A phrase floated in my head to describe Mike and his work— "commonsense genius." I meant no disrespect by the word "commonsense." There is nothing obvious about the concept of obvious, and the ultimate victory for a scientific idea is to become the new common sense, as I have already stressed. I especially admired the confidence that I could place in the results of the beeper studies. Many scientific studies are like a rickety scaffold that might hold but then again might collapse under your weight. As a result, you hesitate to stand upon it to extend the scaffold. The beeper studies were so solid that I could stand upon them with confidence. When Mike and his colleagues concluded that gifted students must enjoy what they are doing on a daily basis to develop their talents over the long term, or that the average couple spends less than ten minutes a day together doing anything that can be called quality time, I *believed* them.

As soon as I had mastered the literature, I wrote a grant proposal to an offshoot of the Templeton Foundation called the Institute for Research on Unlimited Love (IRUL). The title of my proposal was "Altruistic Love, Evolution, and Individual Experience." I would study individual differences in altruistic behavior using the beeper studies of Mike and his colleagues and some new beeper studies of my own. I can assure you that this kind of research would never get done without offbeat foundations to help offbeat scientists like me. The grant was funded and I flew to Chicago to be trained by Mike's colleague Barbara Schneider and her staff, who maintain the computer files of past beeper studies like the monks of old who maintained and transcribed sacred texts. A few days after my return, I received a single CD in the mail containing the results of a beeper study that had required five years and $7 million to complete, completely free of charge. If that's not an act of unlimited love, what is?

33

The Ecology of Good and Evil

THE ORIGINAL PURPOSE OF the beeper study that came in the mail was to answer a down-to-earth question, a bit like Sir John Templeton's "How can we inspire our *young* people?" Mike and his colleagues had obtained a gigantic grant from the Alfred P. Sloan Foundation to study how young people prepare to enter the workforce. Twelve geographical locations in the United States were chosen to represent rural, urban, and suburban environments, different racial and ethnic compositions, labor force characteristics, and economic stability. Within each geographical location, a number of middle schools and high schools participated, resulting in thirty-three schools for the entire study. Over a thousand students took part, not once but three times at two-year intervals, covering their entire middle and high school careers. In addition to being beeped for a week, they also answered hundreds of questions on a one-time basis. As if this was not enough, thousands of additional students answered the one-time questions without being beeped. All of this information was encoded onto a single compact disc, a modern miracle similar to the encoding of life itself in a strand of DNA

As I popped the CD into my laptop and examined the files, I began to appreciate the awesome amount of work required to acquire this sample of American life. Contacting the schools and getting the students to fill out thousands and thousands of booklets over a five-year period was hard enough, but then their responses had to be entered into the computer by hand. Answers to questions on numerical scales could be quickly entered but written responses to questions such as "Where were you?" and "What were you thinking?" had to be organized into hundreds of numerical codes. A team of people had to read handwritten comments such as "arguing with my woman," decide that it should be code 335 ("talking with girlfriend/boyfriend"), and then repeat the exercise ad infinitum. The files on the CD were hundreds of columns wide and thousands of rows long. If I were to print one of them on paper, it would require a large wall to display. A thick spiral-bound book was required to make sense of them, and all of this was just for one of the three sampling periods.

The results of the study had already been published as a book by Mike and Barbara titled *Becoming Adult: How Teenagers Prepare for the World of Work*. My challenge was to use the same information to ask questions about altruism— the who, how, why, where, and when of people helping each other, using evolutionary theory as my guide. There is no single definition of the word "altruism," any more than there is for the word "selfish." The implication of self-sacrifice is foreign to the religious imagination, as we saw in Chapter 29. For the purposes of my study, I focused on the other-oriented dimension of altruism, including anything that contributes to the welfare of others or society as a whole. I often use the word "prosocial" rather than "altruistic" to avoid the implication of necessary self-sacrifice.

Looking over the hundreds of questions that were asked on a one-time basis, I found seventeen that were clearly related to my broad definition of altruism, such as "For the job you expect to have in the future, how important is helping people?" and "Among the friends you hang out with, how important is it to do community work or volunteer?" Each question had been answered on a numerical scale, so the numbers could be combined to provide a single score for each individual. This was easy enough in principle, but actually doing it involved manipulating the files in a way that I found nerve-wracking. Some questions were answered on a scale from one to three, others from one to five, and still others from zero to four. I needed to convert them to a single scale to make them comparable. For some questions such as "Do you think that helping strangers is a waste of time?" a high number indicating agreement implies a low degree of altruism, so these scores had to be reversed. When a student mistakenly marked more than one number for a given question, it was given an arbitrary code of "96," which would appear as a form of super-altruism unless I removed it from the analysis. These modifications were made with computer commands that changed thousands of numbers with a single touch of the enter key. I lived in constant fear of making some colossal mistake and converting all of the numbers into garbage. Worse, I might not realize my mistake and spend the rest of my life searching for patterns that no longer existed! I began to think that obsessive-compulsive disorder is not necessarily a disorder at all but exactly the kind of personality that is required for a task such as this.

At last I managed to combine the seventeen numbers into a single number for each individual, which I called the PRO score, for "prosociality." The distribution of PRO scores described a nice bell-shaped curve, with most students intermediate and a tail of high and low scorers at the ends. Psychologists construct scales like this all the time and have developed methods for assessing their validity. These methods can be quite complex, so I enlisted the help of an expert named Jack Berry, who described what he was going to do this way: Imagine a Latin test that includes both easy and hard questions. Most Latin students will answer the easy questions, but only a few will answer the hard questions. The same will be true for a Greek test that includes both easy and hard questions. Now imagine combining the two tests into a single test. There is no longer a gradient that runs from "easy" to "hard." A Latin student who can answer a hard Latin question will be unable to answer an easy Greek question, and vice versa. The test is invalid because it inappropriately lumps different capacities (in this case knowledge of different languages). Jack assessed my PRO scale and determined that the questions fell along an easy-to-hard gradient. It seemed to be measuring a single capacity (in this case an orientation toward helping others and society), rather than inappropriately lumping different capacities.

My next task was to understand why individuals differ in their PRO scores. It is a scientific mantra that correlation does not imply causation, but correlations do provide a starting point for discovering the web of causation underneath. I therefore combed through the hundreds of questions asked on a one-time basis to find those that correlate most strongly with the PRO score. A statistical technique called multiple regression analysis identified fourteen questions, which collectively accounted for 40 percent of variation in the PRO score and could be grouped into the following categories.

The first category is *gender*. The average male is more prosocial than the average female. Just checking to make sure that you're awake. The average female is more prosocial than the average male, by quite a wide margin.

The second category is *social support*. High-PROs have more teachers who care about them, neighbors who are more likely to help, and families more likely to avoid hurt feelings, compared to low-PROs.

The third category is *self-esteem*. High-PROs are more hopeful about the future, energetically pursue their goals, and feel like a person of worth, compared to low-PROs.

The fourth category is *planning for the future*. High-PROs spend more time on homework after school, think more that it is important to have children and to provide them with opportunities, and expect to encounter obstacles (which they also expect to overcome) compared to low-PROs. On the other hand, low-PROs value immediate gratification such as partying with friends more than high-PROs.

The fifth category is *religion*. High-PROs are more likely to indicate that religion affects their decisions and that religion is important among their friends, compared to low-PROs.

How can these results be interpreted from an evolutionary perspective? Way back in Chapter 5, I described my imaginary desert island thought experiments to illustrate the costs and benefits of traits associated with good and evil, high-PRO and low-PRO. When they are placed together on a single island, the low-PROs have the advantage. When they exist on different islands, the high-PROs have the advantage. When the separation is not complete, there is a messy combination of costs and benefits, with groups of high-PROs doing well by helping each other, despite a degree of exploitation by low-PROs in their midst. The fate of the behavioral strategies depends entirely on how they are clustered.

The results suggest that social interactions in America are quite highly clustered. High-PRO teenagers might *give* more (at least by their own report), but they also *get* more, from teachers, neighbors, and family. Most high-PRO individuals are nestled in the bosom of a high-PRO social environment. Blessed by a stable and nurturing support system, high-PRO individuals are able to develop their potential and plan for long-term goals, rather than watching their backs and worrying about their next meal. Social support, self-esteem, and long-term planning go together as a package.

This is only true in a statistical sense, however, not for each and every individual. The desert island experiments make it clear that a high-PRO individual who finds herself on a low-PRO island is in trouble. Similarly, a low-PRO individual on a high-PRO island can live the life of Riley, at least until he is caught. I should be able to find examples of these outcomes in my sample of American life. Rummaging back through the hundreds of questions asked on a one-time basis, I found some that inquired about significant events that might have occurred during the last two years. Had they moved to a new home? Had their parents gotten divorced or remarried? Had they or a family member been ill? Had they witnessed a violent crime? Had someone threatened to hurt them? So many teenagers were involved in the study that over a hundred answered yes to the question "Have you been shot at during the last two years?"

Each of these questions was followed by a second question for those who responded with a yes: "How stressed were you by the event?" When I correlated the responses to the PRO score, I came up with a fascinating result. High-PRO individuals were less likely to experience adverse events such as being assaulted or witnessing a violent crime. This makes sense because most of them are in the bosom of a high-PRO social environment, as I have already shown. However, when high-PRO individuals *did* experience such events, they were *more stressed* than low-PRO individuals. I call this the "fish out of water" effect. An ounce of evolutionary thinking tells you that *no behavior is advantageous across all social environments*. A high-PRO individual in a low-PRO social environment is in the wrong habitat, like a fish out of water, and has only four options: leave, change her behavior, attempt to change her social environment, or suffer the consequences of being exploited.

In Chapter 5, I was careful to talk about good and evil *behaviors*, not *individuals*. All species can flexibly change their behaviors according to rules laid down by their evolutionary past, and we are the most flexible of all species. This leads to a prediction that I call "more ways to be bad than good." A person who scores low on the PRO scale might be a genuinely self-centered person on the lookout for high-PROs to exploit. On the other hand, he might be a perfectly well-meaning person who is coping with a low-PRO social environment by "turning off" his own prosociality or restricting it to a few trusted associates. To someone who has been abused many times in the past and is struggling to make it on his own, a question such as "For the job you expect to have in the future, how important is working to improve society?" can appear laughably naive. When it comes to low-PRO behaviors (or any behavior), we must evaluate the social environment at least as much as the individual. When Leo Tolstoy began his novel *Anna Karenina* with the sentence "Happy families are all alike; every unhappy family is unhappy in its own way," he meant that happiness requires a number of ingredients and the absence of any one of them can cause the recipe to fail. As for happy families, so also for prosociality in general.

To test my "more ways to be bad than good" hypothesis, I

compared the upper and lower tails of the PRO distribution. Sure enough, the low-PROs were more variable than the high-PROs in their response to dozens of one-time questions relevant to self-esteem, emotional emptiness, long-term goals, efficacy, trust, deviance, and stress. The low-PROs were a heterogeneous mix who had achieved—or been reduced to—their low prosociality in different ways. The high-PROs were more homogenous, except when it came to religion. For seven questions that were asked about religion, high-PROs were more variable than low-PROs for each and every one. If you are a low-PRO, you are probably not religious. If you are a high-PRO, you might or might not be religious. I am a good example of a nonreligious high-PRO, for example.

As a group, low-PROs give the appearance of being a downtrodden bunch. They suffer from low self-esteem, are pessimistic about the future, and believe that luck is more important than hard work. They score significantly higher on the items "I usually feel stressed," "I usually feel sick," and "I usually feel tired." This profile is very different from the image of an egoist who is full of himself (high self-esteem), is planning for a house in the Hamptons (optimistic about the future), and leaves nothing to chance (believes that hard work is more important than luck). The reason that puffed-up egoists are not more conspicuous in the regression analysis is because they constitute only part of the heterogeneous mix of low-PROs. As soon as I began distinguishing among different *types* of low-PROs with a statistical technique called cluster analysis, the puffed-up egoists emerged in all their glory. My choice of the masculine gender is appropriate because 80 percent of them are male.

I was also able to use cluster analysis to explore differences between religious and nonreligious high-PROs. Despite their shared prosocial values, religious high-PROs have higher self-esteem, appear to work harder toward their future, and submit to more social control (concerning such things as household chores and limits to watching TV), but they actually feel that they have *more* control over their lives than nonreligious high-PROs. These results affirm the practical benefits of religion that I also stressed in Chapters 28 and 29.

All of these results are based on the hundreds of questions that the students answered on a one-time basis. Do high-PRO and low-PRO individuals also respond differently when they are beeped at random during their daily lives? Indeed they do. High-PROs report that they are concentrating better; that they are living up to the expectations of themselves and others; that they feel better about themselves; that they are happier, more active, social, involved, and excited; that they are more challenged by activities that are more important and difficult; that what they are doing is more interesting and relevant to their future goals; and that they are making better use of their time. There is not a shadow of a doubt that if you are an American teenager, being a high-PRO is hugely benefi-cial, but only if you are in the bosom of a high-PRO social en-vironment. These are results that you can stand upon, thanks to the enormous amount of work accomplished by Mike, Barbara, and their collaborators, the financial backing of insti-tutions such as the Alfred P. Sloan Foundation and the John Templeton Foundation, and the beauty of the beeper method.

These results are described in more detail in a paper writ-ten with Mike titled "Health and the Ecology of Altruism," which is available on my Web site. Much more remains to be discovered, but I am doubly happy with our analysis so far. In the first place, it shows how people can be studied in exactly the same way that ecologists study other species in their natural environments. In the second place, the most basic evolutionary predictions about altruism as a social strategy that can succeed or fail, depending upon the social environ-ment, are proving their worth—not only for lofty subjects such as the nature of religion but also for down-to-earth sub-jects such as the daily experience of American youth. It is not yet common to think of human behavioral diversity as like biological diversity, but it might just become the new com-mon sense.

34

Mosquitoes Under the Bed

THE PANAMA CANAL WAS the largest engineering project of its time and it almost failed because of yellow fever and
malaria. In their ignorance, the French and Americans attributed these diseases to moral weakness. According to David
McCullough in *The Path Between the Seas*, "Nearly everyone
was profoundly shaken whenever the death of some notably
upright person seemed to make a mockery of such views.
'Certainly his moral quality was above reproach,' wrote one
bewildered, grieving French engineer of another who had
died of yellow fever the first year." Eventually it was discovered that both diseases are transmitted by mosquitoes. In
fact, mosquitoes were breeding under the very beds of the
hospital patients, in bowls of water that were placed under
each leg to deter ants. Once the true causes were understood,
practical measures could be taken to bring the diseases under
control.

The message of this story is that some problems require
practical solutions. All the moralizing in the world couldn't
prevent yellow fever and malaria. An ounce of factual knowledge could.

We need factual knowledge today more than ever. In addition to our continuing battle against disease organisms, we

must battle new problems of our own creation, from compounds in plastic that mimic hormones to global climate change. In each case there is something comparable to mosquitoes under the bed that has nothing to do with moral weakness and must be discovered to achieve a practical solution. Yet it would be naive to portray factual knowledge as unambiguously good. It is the cause of our problems as much as the solution. In Kurt Vonnegut's novel *Cat's Cradle*, an ounce of factual knowledge in the form of a newly discovered stable configuration of water combines with human folly to bring about the end of the world.

Like it or not, the Pandora's box of modern technology has already been opened and we must deal with the consequences. As this book draws to a close, it is worth reflecting in general terms upon science, moral values, and evolution.

At least three problems make factual knowledge a mixed blessing. The first is *unforeseen consequences*. Plastic appeared unambiguously good when it was invented, and nobody had the slightest idea that it might release hormone mimics into the environment. The second is *unethical use*. Facts are powerful and can be used as weapons by some against others in addition to their more benign uses. The third is the *erosion of moral values*. It's easy to smile at the idea of malaria as a moral weakness, but some problems *are* caused by moral weakness and solved by implementing a strong moral community, as we have seen in previous chapters.

These problems might be enormously difficult to solve in practice, but the way to begin is relatively straightforward, at least as far as I can see. To solve the problem of unforeseen consequences, we need to be suitably humble about what we know, cautious about implementing new technologies, and diligent about discovering unforeseen consequences. The ultimate solution to partial knowledge is more complete knowledge.

To solve the problem of unethical use, we need ethical social systems that prevent the exploitation of some by others. There is nothing special about factual knowledge per se when it comes to ethical conduct. If we can create ethical social systems, then the use of factual knowledge will become more benign along with everything else.

To prevent the erosion of moral values, we must think carefully about the relationship between practical and factual realism. Is there *necessarily* a trade-off between beliefs that enable people to flourish in sustainable communities and factual knowledge? We know about trade-offs for particular belief systems, but is there a more general trade-off that can't be avoided? It was upsetting to learn that the earth is not the center of the universe, but the discovery did not permanently damage our moral systems as far as I know. A more general trade-off would pose a more serious dilemma.

Evolutionary theory is deeply relevant to each of these three problems and their solutions. Let's begin by asking if the theory leads to factual knowledge that can be used to solve the practical problems of life. If I were to pose this question to a biologist, you can imagine the incredulous look that I would receive in return. The mantra "Nothing in biology makes sense except in the light of evolution" was coined over thirty years ago. Yet if I were to pose the same question to people in a bar or a supermarket, I would receive another kind of incredulous look. A few might mumble something about antibiotic resistance but the vast majority wouldn't have the slightest idea how to reply. The most extraordinary fact about public awareness of evolution is not that 50 percent don't believe the theory but that nearly 100 percent haven't connected it to anything of importance in their lives.

If this book has had any impact on your thinking, I hope it is to demonstrate the enormous relevance of evolutionary theory to matters of importance in human affairs. To pick just one of many examples from previous chapters, we along with other mammals are designed to assess our nutritional environment very early in life and adopt a metabolic strategy that will last the rest of our life. This "predictive adaptive response," or PAR, has become maladaptive in modern environments (an example of "dancing with ghosts") with medical complications that cause as much misery and death as malaria or yellow fever. It was not discovered until very recently because it is impossible to even imagine unless you're thinking like an evolutionist. This and many other discoveries based on evolutionary thinking are purely beneficial and essential for solving problems comparable to mosquitoes under the bed.

The reason that we believe so firmly in the physical sciences is not because they are better documented than evolution but because they are so essential to our everyday lives. We can't build bridges, drive cars, or fly airplanes without them. In my opinion, evolutionary theory will prove just as essential to our welfare and we will wonder in retrospect how we lived in ignorance for so long. If you think I am exaggerating, remember my metaphor of a person trying to explain a sculpture without any concept of an artist. Now consider that the sculpture is the entire living world, including our own species. Knowledge of the "artist" is essential, and we are fortunate that her "intentions" are discernable, in contrast to the inscrutable intentions of supernatural agents and other "intelligent designers" that have been imagined over the decades and millennia. To summarize, if we value factual knowledge at all, we must place a very high value on evolutionary theory and work hard to expand it beyond the boundaries of the biological sciences, where it has been restricted for most of the twentieth century.

With respect to the ethical use of factual knowledge, evolutionary theory has often been misused, as I was careful to acknowledge in Chapter 2. Is it *more* prone to misuse than other scientific theories, however? A fascinating book by Rebecca Lemov titled *World as Laboratory: Experiments with Mice, Mazes, and Men* recounts the history of the social sciences in America, including the "blank slate" tradition of behaviorism. Behaviorism was also misused, suggesting that the real problem is how to avoid the misuse of *any* theory, not just evolutionary theory.

In retrospect, the early social sciences in America are striking for their grandiose expectations, as if full understanding and control of man and nature were just around the corner. Jacques Loeb, who provided the model for the scientist-hero of Sinclair Lewis's 1925 novel *Arrowsmith*, told a reporter, "I wanted to take life in my hands and play with it—to start it, stop it, vary it, study it under every condition." Loeb's protégé, John B. Watson, echoed the same sentiment: "I believe we can write a psychology and ... never use the terms consciousness, mental states, mind, content, introspectively verifiable, and the like.... It can be done in terms of stimulus and response.... My final reason for this is to learn general and particular methods by which I may control behavior." Continuing the tradition, Clark Hull achieved prominence not for his scientific contributions (which have been completely forgotten) but by acting as a kind of a prophet for the malleability and control of behavior. As Lemov puts it, "he presented his ultra-behaviorism as salvation." Hull's radical behaviorism made evolution appear irrelevant. If behavior is *that* malleable, who cares what happened during the Stone Age?

These grandiose expectations were combined with a blind faith in authority, as if scientists, politicians, and the captains of industry could be expected to do what's right without any oversight. Lemov's account of the Laura Spelman Rockefeller Memorial Foundation, which empowered Beardsley Ruml to almost single-handedly define the social sciences, captures the spirit of a bygone age:

> The memorial's policy papers from these years (1922–1926) used the term "Social Welfare" pretty much synonymously with "Social Engineering" and "Social Intelligence" and "Social Technology." Once the new social science had explored and mapped the human and social realm in a properly replicable and as-objective-as-possible manner, change would necessarily follow. Under the banner of scientific reason, even the irrational elements of society were susceptible to control. Crime, delinquency, and abnormal sexual or familial function could be corrected through the redesign of environmental and social relations; perhaps scientists could even address unbelief and the ravages of twentieth-century normlessness—what Ezra Pound had called in 1915 "a botched civilization," in which crowds, masses,

entire populations inhabited a world without much sense or sensibility—by substituting new norms of behavior and alternative forms of faith. Controlling social technology, it followed logically, was a task for social scientists and other experts, for who was in a better position to render knowledge as technique? Faced with the monumentality of their task, human and social engineers reminded themselves that they were mere servants or technocrats working for those who set forth the ultimate goals of social control: democratically elected officials. Without social-control experts, such leaders, short of adopting authoritarian methods, could hope for little effect. Thus social sciences were the greatest hope for democratic social control: they were hope itself.

Finally, grandiose expectations and blind trust in authority were combined with a willingness to inflict suffering on individuals to benefit society as a whole that appears shocking to us today. The worst excesses took place during World War II and the Korean War under the guise of national security. Harold George Wolff, a leading authority on pain who conducted covert research on mind control funded by the CIA, boasted that he could provide the keys to "how a man can be made to think, feel, and behave according to the wishes of others, and, conversely, how a man can avoid being influenced in this manner." Some of the experiments conducted during the Korean War by leading scientific authorities, involving ice-pick lobotomies that leave no scar or the total breakdown of a person prior to "reconstruction," are as ghoulish as anything that can be imagined in fiction or the dark recesses of a totalitarian state. Most social scientists were unaware of these experiments and were willing to receive military funding for their own research. As one CIA scientist put it during a 1977 congressional hearing, "Don't get the idea that all these behavioral scientists were nice and pure, that they didn't want to change anything and that they were detached in their science. They were up to their necks in changing people. It just happened that the things they were interested in were not always the same as what we were."

Lemov's gallery of social scientists concludes with Timothy Leary, who had rather different ideas about how to alter human behavior. As a young and soon-to-be-fired Harvard professor in 1961, he obtained permission to "turn on" with inmates of a local prison to break down the barrier between scientist and subject. Leary recalls that his first trip with a Polish embezzler named John got off to a rocky start.

> Doc, he said, why are you afraid of me? I said, I'm afraid of you, John, because you're a criminal. He nodded. I said, John, why are you afraid of me? He said, I'm afraid of you, Doc, because you're a mad scientist. Then our retinas locked and I slid down into the tunnel of his eyes, and I could feel him walking around in my skull and we both began to laugh. And there it was, that dark moment of fear and mistrust, which could have changed in a second to hatred and terror. But we made the love connection. The flicker in the dark. Suddenly, the sun came out in the room and I felt great and I knew he did too.

Leary's treatment had a profound effect on the prisoners, but only for as long as their relationship lasted. Even though his "experiment" can hardly be called scientific, it hints at a vital ingredient for the ethical application of knowledge—mutual trust. People dread being controlled by others without their consent, for the best of reasons. When people agree upon their social priorities by consensus, then the practical application of scientific knowledge becomes benign. When knowledge is used without mutual consent, it becomes sinister. It doesn't matter whether the knowledge is based on evolutionary theory or its polar opposite, radical behaviorism.

The question then becomes: how can we build bonds of mutual trust more durable than Timothy Leary's drug-induced state? Evolutionary theory has much to say about this subject, as I have recounted in previous chapters. The social sciences are full of scenarios about the lives and minds of our ancestors prior to the advent of civilization. Rousseau imagined a noble savage corrupted by society. Hobbes imagined a brutish savage that must be tamed by society. Freud imagined a guilty savage whose patricidal act somehow became embedded in racial memory. Economists imagine a selfish savage, sometimes even referred to as *Homo economicus,* who becomes civilized only by appealing to his self-interest. It is worth asking why these origin myths are necessary when they have no more basis in fact than the Garden of Eden. My guess is that they play a practical role in the belief systems that create and sustain them, much as the distorted versions of history in the four Gospels. In any case, we are on the verge of replacing these scientific creation myths with more authentic knowledge about our species as a product of genetic and cultural evolution. This knowledge can almost certainly help us build bonds of mutual trust, which in turn can promote the benign application of all knowledge by mutual consent.

Finally, there is the question of whether a belief system can combine the best of religion and science, enabling people to flourish in sustainable communities while remaining fully committed to factual realism. It is important to realize that this would be a new cultural adaptation, never before seen on the face of the earth. Factual realism has always been the servant of practical realism, showing up when useful and excusing itself otherwise. This has been true starting with the perceptual systems of bacteria. Our minds are genetically designed to encode instructions for how to behave as factual statements. It is as natural for us as having sex and demonizing our enemies. Only now, in highly differentiated modern societies, has it become important to create a large body of factual knowledge that can be trusted, to solve practical problems at an unprecedented social, spatial, and temporal scale. Fortunately, human moral systems are flexible enough to embody anything that is deemed good and right, even if it demands discipline and self-restraint, as it usually does. The first step is to decide that factual knowledge is a virtue—

sacred, if you like—and that value systems must treat statements of fact more respectfully in the future than the past. His Holiness the Dalai Lama points the way when he states that "to defy the authority of empirical evidence is to disqualify oneself as someone worthy of critical engagement in a dialogue."

35

The Return of the Amateur Scientist

SCIENCE IS SUPPOSED TO be the supreme intellectual activity, beyond the reach of all but an elite few. Even they must train for years beyond their college education to acquire their advanced skills, becoming so specialized that only a handful of peers can speak their language and comprehend the significance of their work.

As if to affirm the superiority of scientists, the general population seems to get dumber and dumber. The nation's children struggle to read and write, avoid science courses like the plague, and take more interest in their social lives than anything that can be called intellectual. The nation's adults work mind-numbing jobs and are distracted in their leisure hours by all manner of diversions, most of them offering short-term pleasure without long-term gain.

It's hard to imagine that only 150 years ago, science was largely an activity that amateurs performed in their spare time. Some were wealthy gentlemen, but others were parish priests, schoolteachers, or doctors. Alfred Russel Wallace, who independently derived the principle of natural selection, provides a nice contrast to Darwin the country gentleman. He had to scramble for his livelihood and became interested in the natural history of his surroundings when he took up the profession of surveyor. Beatrix Potter, who is beloved today

for her children's books, was a busy sociologist and social reformer who also found time to be an accomplished mycologist and was the first to discover that lichens are a symbiotic
association rather than a single species.

It's easy to conclude that those days are gone forever and
that science has become restricted to the narrowly specialized professional scientist. Yet this book tells a different story.
I and my evolutionist brethren have escaped the fate of narrow specialization. Even those who specialize, such as Doug
Emlen with his dung beetles and Tom Seeley with his honeybees, address fundamental issues that can be communicated
to the rest of us without complicated jargon. Years of training
are not required. A single course is sufficient to get my students started, and this book was written to provide the same
service at a distance. The club is not restricted to elites. I work
at a state university that doesn't require a fortune to attend.
My students come from all ethnic groups, all walks of life,
and increasingly from around the world. It's true that they
needed to excel at their studies to get here, but I don't think
that one's location on the academic bell curve is determined
at birth. Innate differences do exist, but regardless of the
genes that we possess, our trajectory through life is more like
a ball in a pinball machine than a straight shot. I have enjoyed
asking my students and colleagues about how they became
scientists in the process of writing this book. They came from
everywhere, male and female, blue- and white-collar, on the
basis of a book they picked up in a used bookstore, a patch of
ground next to their house, or their attachment to a pet parakeet. Their advancement appears to be based more on their
motives and the pleasure they get from what they are doing
than raw intelligence.

Perhaps it is fitting for me to end this book by describing
my own early trajectory through the pinball machine of life.
It is not a rags-to-riches story, but it does reflect another
transformation that can help us to understand the difference
between science and non-science, professional and amateur,
smart and dumb. I mentioned earlier that my father was
Sloan Wilson, whose novels included two blockbusters, *The
Man in the Gray Flannel Suit* and *A Summer Place*. The first

became an icon for the fifties generation, and the second became an icon for changing sexual mores. Both were made into successful movies and you can still hear the theme to *A Summer Place* on the radio.

The Man in the Gray Flannel Suit was published in 1955, when I was six years old. Before then, my father had published stories in magazines such as the *New Yorker* and a wartime novel titled *Voyage to Somewhere*, but he couldn't even remotely survive as a professional author. A video documentary titled *The Fifties*, based on David Halberstam's book of the same name, includes an interview with my father describing this period of his life. He was working as an assistant to Henry R. Luce, head of *Time* magazine and one of the most powerful men in America. One day, another assistant tried to curry favor with Mr. Luce by offering to hold his hat. My father, who had captained supply ships during the war, suddenly realized that he had been reduced to the role of a butler. "May *I* hold my man's hat?" he asked plaintively, and no one noticed the irony of the request. He quit his job the next day.

His next job was even more lowly. He served on the public relations staff at the University of Buffalo, organizing social functions and giving tours of the campus to the parents of prospective students. On one tour of a woman's dormitory, a mother asked if her daughter would be safe from the sexual advances of the male students (remember that this was the 1950s). No sooner had my father reassured her than he looked down and spied a used condom on the floor. What does a public relations man do? He puts his foot on it, of course!

Given such an unrewarding job, the main thing that gave life meaning for my father was his writing. It didn't matter that he came home tired at the end of the day to a wife and three children. He still headed down to his study in the basement to beat away at his typewriter at the first opportunity. I dimly remember falling asleep to the sound of his typewriter, the way that other children fall asleep to the sound of their furnace.

My father's passion for finding meaning through writing is

not unusual. For every successful author, there are hundreds of would-be authors pounding away at their typewriters or computers. Some are driven by a desire for fame and fortune but others are driven primarily by the desire to find meaning. For every would-be author, there are thousands of people who write in journals purely for its own reward, without any thought of fame or fortune. The search for meaning is a primal human desire, as the great psychiatrist and Holocaust survivor Viktor E. Frankl emphasized in his work. You're lucky if you can get paid for it, but otherwise it is something that you pay to do. That is the main difference between a professional and an amateur when it comes to writing.

My father was one of the lucky few who could support himself on his writing. Not only did *The Man in the Gray Flannel Suit* change his life, but it also changed mine. Try to imagine what it is like for a small boy to have a father who is the center of attention wherever he goes. At restaurants he was approached by young women seeking his autograph. The local aristocracy invited him to their parties, and I was sometimes allowed to tag along. We would drive up to a mansion on manicured grounds far above our station. The other guests would be as deferential as the servants, but somehow my father outranked even the hosts. Whatever he had trumped money and pedigree.

My father was a king, which must have made me a prince, but I didn't feel like one. I didn't even look like my father. He was a lion of a man with piercing gray eyes and hair swept back in a mane. I looked more like a slender gazelle. It's a quirk of genetics that men sometimes resemble their maternal grandfathers more than their fathers. I physically resembled my mother's father, a traveling salesman who sold intimate apparel before marrying into a wealthy Boston family.

Self-analysis is treacherous, but when I look back upon my boyhood, it seems that everything I did was designed to avoid direct comparison with my father. He loved boats and was a champion sailor as a boy on Lake George in the Adirondack Mountains, where his family used to run a hotel. We still summered on the grounds of the hotel, which was dismantled during World War II but rebuilt by my father's imagination in

A Summer Place. As soon as I was old enough, my father bought me a little red sailboat. You'd think that it would be the perfect gift for a boy and an opportunity to be with his father, but the stress of attempting to equal something that he did so well was unbearable for me. On the other hand, when I was given a kayak a year later, which had two seats but was too narrow for my portly father to fit into, it became my constant companion. I also developed a passion for fishing, for which my father had no patience. I wonder what he thought as he watched me tirelessly patrol the edges of the great wooden pier, which long ago had received the hotel guests arriving by a paddleboat that still churned past as a tourist attraction, sending its wake rolling onto our shore. I had made myself literally unapproachable, lest he scare the fish.

Even at a young age, I was contemplating life's options with a seriousness not usually associated with childhood. Playing with and proving myself among my peers had little appeal. I had to fill my father's shoes, but the prospect of equaling him at what he did so well was too daunting. I needed to do something that he would admire but couldn't do himself, preferably something that he couldn't even evaluate. I would become a scientist. I didn't know what this meant. All I knew was that when talk turned to what I wanted to do when I grew up and I said something like "brain surgeon," my parents' eyes would light up.

Unbeknownst to me, my parents were on the verge of divorce, and I was sent to boarding school at the tender age of eleven. My mother later told me that I was following my father around saying, "I'm sorry, I'm sorry," which struck her as unhealthy, although I don't remember this myself. They sent me to the North Country School near Lake Placid, New York, not far from our summer place on Lake George. It is still thriving and I recommend it to anyone who can afford to go, regardless of the quality of their home life. It offers the healthful outdoor life of an actual working farm, and it is this that I remember more than my academic studies. We collected the still-warm eggs and even slaughtered the chickens and pigs that we ate for dinner. On winter mornings so cold that it froze the hairs in my nose, I stood on the giant compost pile and received the new steaming manure from the "honey cart" that ran past the cow and horse stalls to build the pile even higher. In the evening I mixed pig mash with the day's garbage and dumped it into the trough, stepping back quickly to let the pigs rush in with unadulterated joy. Then I would jump on one, rodeo style, and hang on laughing until it shook me off its hard, bristly body. How strange, that my family had to be rich to provide me with the life of a farm boy.

After graduating from the North Country School in the eighth grade, I tried to find another just like it in the form of the Woodstock Country School near Woodstock, Vermont. Unfortunately, Woodstock was inhabited by hormone-crazed teenagers instead of sweet little children, and that made a big difference. Moreover, the permissive philosophy of the school placed it in the direct path of the sex, drugs, and rock-and-roll hurricane of the sixties. All except for me. I was on my way toward becoming a great scientist. Because I was one of the only students with academic ambitions, my grateful biology teacher allowed me to use a little room adjoining the classroom as a laboratory. My own laboratory! I strutted around in a white coat and fiddled with the ancient equipment that was stored there. Somewhere I read about an experiment where rats were trained to expect food at the sound of a high tone and water at the sound of a low tone. Gradually the high tone was lowered and the low tone raised until their meaning became ambiguous, which supposedly made the rats go crazy. I wanted to repeat the experiment and connected an electric air pump to a harmonica to create the high and low tones. Unfortunately, even at the pump's lowest setting the harmonica made a piercing wail that could be heard throughout the campus. This experiment required sacrificing the rats at the end and weighing their adrenal glands as a measure of stress. The school did not have adequate dissecting tools, so my teacher purchased a lovely pair of stainless steel surgical scissors. I will never know how he meant it—perhaps only as a needed piece of equipment for his classroom—but for me it was like a personal gift when he presented it to me in its elegant felt-lined protective case.

I don't know what my schoolmates thought of me, but I received an inkling from a girl who shared my interest in biology and upon whom I had a terrible crush. She compared me with two other boys this way: One, who was blond and made her laugh, she described as day. Another, who was swarthy and liked to walk on the wild side, she described as night. "And you, David," she concluded, laughing at her own cleverness, "are dusk."

In my senior year I applied to the very best Ivy League colleges and was shocked when they turned me down. I ended up attending the University of Rochester, which I had regarded as my "safety" college. Its curriculum was just as rigorous as my top choices, even though it lacked the prestige that I craved. Somehow, blowing air through a harmonica and the rest of my progressive education had not prepared me for the tough science courses and rows of students bent on becoming doctors. Physics, chemistry, biology, and especially calculus were hard to learn and easy to forget. Even after giving it my very best shot, my average grade was a miserable C+ at the end of the first semester. Only by a massive act of self-deception was I able to cling to my dream of making a mark upon the world.

I was horribly lonely and, when I finally managed to meet a girl named Mary in the library, I pursued her like a starving aborigine lunging after a rabbit. She was a "townie" taking a night class and she already had a boyfriend, but nothing could stop me. Once during the spring, when she was about to drive away with her boyfriend in his car, I ripped an entire branch from a lilac tree in bloom and threw it into the backseat. Later she told me that this impetuous act, combined with the heady scent of lilac filling the car, finally won her heart.

That summer I had a job as a stockroom assistant at the Marine Biological Laboratory (MBL) in Woods Hole, Massachusetts. Woods Hole is a picturesque town on Cape Cod that hosts two world-renowned scientific establishments: the Woods Hole Oceanographic Institute (WHOI) in addition to the MBL. Every summer the town is flooded with scientists and graduate students who come to work hard and play hard. It was hard to tell the difference between work and play for these people. Long hours in the laboratory and classrooms were interspersed with time on the beach or in the bars that lined the pretty harbor dotted with sailboats, but always the talk was about science. They never stopped talking about it because nothing else was as interesting for them. Apart from my father, I had never seen such driven people, even though I couldn't understand what they were so excited about.

I was the stockroom assistant for the physiology course, which was taught by a dozen eminent scientists to about thirty elite graduate students. I was paid to keep track of the equipment and supplies but I was there to learn so I sat in on the classes whenever possible. These people were busy taking life apart to see how it works. The more you take life apart, the smaller the pieces become, until finally you can't even see or manipulate the parts without the help of complicated instruments. Years of hard work are required to figure out how the parts interact at this minute scale, such as how muscle fibers slide along each other to contract or how calcium is pumped across nerve membranes. I found the classes interesting but difficult to understand, just like my college courses. Then by chance I happened to walk past the marine ecology class in the adjacent building. The professor who was lecturing didn't look like the physiologists and didn't wear a lab coat. In fact he looked like Walt Whitman and was wearing lederhosen. He was talking about animals that I could see and how they interacted with each other and their environment. I suddenly realized that I could be this kind of scientist and do the things outdoors that I already did for the love of it. Right away I decided to be an ecologist.

When my job ended in late July, I took Mary to meet my father and his second wife, Betty, who were summering on Lake George. I have never seen two people grow to hate each other as quickly as my father and Mary. She smoked cigarettes and he occasionally smoked cigars in addition to his pipes. On their very first meeting he lit a cigar, and Mary said, perhaps as a lame attempt at humor, "I hate cigars—they're nothing but a big penis symbol." Never say something like that to my father. He smiled malignantly and replied, "Well, what is a cigarette but a *little* penis symbol?"

Continuing the life-as-pinball-machine metaphor, for the next few days I felt like that silver ball being batted between two electrified bumpers. I literally went back and forth between my father and Mary, because they avoided being with each other. Something had to give, so instead of spending the month of August at Lake George, Mary and I climbed into my beloved father-proof kayak and paddled 400 miles back to college, first down the Champlain Canal to Albany and then the Erie Canal to Rochester, camping along the shore or staying in motels in towns along the way.

Back at Rochester, I immediately acted upon my new ambition to become an ecologist by taking the ecology course offered by professor Conrad Istock. I could relate to some of the material but much of it was hard to learn and easy to forget, just like my other science courses. I was even surprised to encounter mathematical equations that were supposed to represent how animals interacted with each other and their environment. Fortunately, the course included an option for me to do an independent project of my own choice. This was more like what I was used to at Woodstock. I gave every spare moment to my project. Somehow I had become interested in the fact that zooplankton—the minute animals that live in open water—migrate vertically during the course of the day, surfacing by night and descending by morning's light. I decided to study vertical migration in the river that flowed by campus and had a great time inventing ways to collect zooplankton from different depths with a garden hose and hand pump. I was also taking a computer science course. In those days computers filled entire rooms and programs were written on stacks of punch cards that had to be submitted to a staff of attendants, only to be returned with the inevitable errors. The computer science class also required a project so I wrote a computer simulation of vertical migration, which did little more than make imaginary plankton go up and down. Looking back, I can imagine Conrad smiling at my wealth of desire and poverty of ability as he read my report, at least five or six times thicker than the others.

Despite my low grades and lack of polish, Conrad decided to accept me into the Biology Department's honor program, which meant that I could work in his laboratory and take whole courses based on independent study rather than lectures. Now I was beginning to find my balance. The books and articles that scientists wrote for each other were more interesting than the textbooks written for students, even if portions went over my head. Helping Conrad with his own research on a species of mosquito that lives in the water-filled leaves of the carnivorous pitcher plant took us into bogs, which are lakes covered with floating vegetation. In a good bog the ground actually ripples under your feet, as if you were walking on the surface of a giant waterbed. Sometimes you can fall through, in which case you must remember to hold your arms out! After visiting the bogs we would often stop at Conrad's house on the way back to campus for a bottle of beer and relaxed conversation. These facets of science were easy to enjoy.

I continued working on vertical migration in zooplankton with minimal supervision from Conrad. I built another contraption for sampling water at different depths. I dreamt of publishing a description of it and having other scientists use it. For me this would have been as thrilling as writing a best-selling novel. My kayak was now my research vessel from which I tested my water sampler. Alas, it was a Rube Goldberg device that looked humorously like an underwater accordion and would never work well. At one point I discovered that if you place a single zooplankton in a drop of water on waxed paper, the drop forms into a bead, like a minuscule fishbowl with the zooplankton inside. If you also put algae into the water, the filtering appendages of the zooplankton make the algae swirl around the drop like batter through an eggbeater. I thought I could use this method to study the zooplankton's feeding behavior without realizing that its watery prison was so small that nothing it did could be trusted to occur in its wide-open natural environment. Nothing I did in Conrad's laboratory came even remotely close to a scientific publication, so my dreams of becoming an author within my own domain remained unfulfilled.

Of the many scientists trying to explain why zooplankton vertically migrate, one proposed that the adults descend into the food-poor depths to avoid competing with their young, who remain at the food-rich surface. Others pointed out that this would be a form of altruism, since the adults would be providing food for everyone's offspring and not just their own. This was my first inkling that a behavior such as altruism, which seems so uniquely human, might also be practiced by a creature as small and humble as a zooplankton.

My relationship with Mary grew steadily worse, but it was my conviction that any relationship could be made to work with enough effort. We had long phone conversations during which she became increasingly upset, the way that Ravel's *Bolero* gets louder and louder. Then she would cheer up and sign off as if some kind of catharsis had taken place, leaving me completely drained. My roommate, a humorous and easygoing guy, overheard these conversations and finally gave me some advice.

"David," he began, "I admire your strength in maintaining this relationship. You are like a strong man in the circus holding an elephant above your head. But, David—*the elephant is shitting on you!*" Alas, even this lucid analysis did not alter my conviction that relationships could be maintained by sheer force of will.

When it came time to apply to graduate school, I again applied to the best in the land, only to be rejected by every one but my "safety," Michigan State University. My relationship with Mary was becoming more labored by the day, but we still decided to get married. My entire family, including my father, put on a brave face for the wedding, which took place in the college chapel on the same weekend as my graduation.

During the summer before starting graduate school, I returned with Mary to Woods Hole, this time as a student in the marine ecology course that decided my fate three years before. I did a project that involved feeding zooplankton colored plastic beads the size of algae, this time in a suitably sized container. Certain sizes of beads would be added to the water and after a few minutes of feeding the zooplankton would be sacrificed and mounted on microscope slides.

Viewed through a microscope, the beads that the zooplankton had eaten could be seen and measured through their body walls and compared to the sizes of beads they had encountered. I can still see those clusters of brightly colored beads inside the stomachs of the zooplankton through the microscope, like so many clusters of children's balloons. The results clearly showed that the zooplankton were not feeding randomly but were selecting the largest beads. When I gave my presentation at the end of the class, one scientist in attendance was Ed Wilson, whose name has been a thread running through the entire tapestry of this book. At the end of my talk he raised his hand and said, "That's new, isn't it?" These four words shot me out of a cannon of elation, and I celebrated that night by getting drunk. A fact! I had documented my first fact! This study became my first scientific publication and really did mean as much to me as a best-selling novel.

The apartment complex for young married couples at Michigan State was called Spartan Village and lived up to its name. Not only were the buildings barrackslike, but they were so numerous and similar to each other that I was forever getting lost on my way home. Our apartment was right next to a bank of four railroad tracks connecting Chicago to Detroit. Every train became deafeningly loud before receding into the distance, but exactly how loud depended upon whether it ran on the track directly underneath our window or the farthest track twenty yards away. Nevertheless, despite our Spartan existence, Michigan State turned out to be a good choice for me, if not for Mary. The ecology graduate program was small but lively and encouraged independence, which was never a problem for me. I started building another contraption, this time to sample the horizontal distribution of zooplankton. I had more dreams of having others use my device—the Wilson sampler—even though the total number of people on the planet who might be interested could fit into a single rowboat. A man named Jim who ran the stockroom in the biology department was also a good builder and in seeking his help we became friends. Jim was old enough to be my father but had a stability and unquestioning nature that was the polar opposite of my father. As a boy he had misbehaved and was told that he might go to the bad place. He actually was convinced by this gentle admonition and became deeply religious for the rest of his life. As a younger man he did missionary work in Africa and told me marvelous stories of his experiences. One day he asked if he and a friend could visit me in my apartment to talk about the scriptures. I couldn't refuse and spent an awkward Saturday afternoon with them around my kitchen table. Then he asked if I would like to go with him to a big revival that was coming to town. Again I couldn't refuse and found myself in a cavernous auditorium with organ music and thunderous preachers. The climax came at the end when people were supposed to walk to the front to be born again. I was astonished when people all around me rose and filled the aisles. I glanced at Jim from the corner of my eye to see him glancing at me, but nothing could have been so foreign to my experience or my own way of thought.

Mary had absolutely nothing to do in Spartan Village and ended up getting a job as a sales clerk at the local Kmart. I would pick her up in the evening and get a soggy submarine sandwich at a bargain price for my next day's lunch. When she was assigned to the camera counter she developed an interest in photography. At first it was just a hobby, but soon she was looking at job ads and actually landed a job at a tiny advertising company. Here was a world that would allow her to expand, just as I was expanding into science. Within a year the frayed rope of our relationship snapped and we filed for divorce. She has since become a successful professional photographer. It was a huge relief to go our separate ways, but it also left me alone except for a few graduate student friends. What was I supposed to do—approach a woman in a bar with the line "Hi! My name is David and I spend every spare moment studying zooplankton"? It seemed that I would forever remain dusk in the eyes of women, but all I could do was keep on trudging.

One opportunity available to me was a course in tropical ecology offered by a consortium of universities called the Organization for Tropical Studies (OTS). Graduate students from the member universities traveled with a team of professors around the tropical paradise of Costa Rica, visiting habitats as diverse as lowland tropical rain forests dripping with moisture, high-altitude cloud forests shrouded in mist, regions with pronounced dry seasons baked by the sun, and island beaches with their surrounding coral reefs. The course lasted two months and the stay at any particular location lasted from a few days to two weeks. Most of the locations were biological stations with resident scientists who gave lectures and led field trips along with the team of professors who accompanied the students. It was like a traveling version of Woods Hole.

I applied to the course and was accepted for the winter of my second year in graduate school. It was a good time to escape bleak East Lansing, made bleaker by my loneliness and the divorce proceedings. The students and teachers converged from all across America to a motel in San José, Costa Rica's capital city, to meet and then travel to the first field site the following day. All my senses were on high alert as I attended the first social gathering at the motel. Not only was the location exotic, with lizards skittering among big fallen banana leaves and the fruity smell of a nearby brewery hanging in the humid air, but I was about to meet the people with whom I would be spending the next two months. Far from stereotypical scientific nerds, most were outgoing and some of the women were very attractive, as I noticed before anything else. I was too shy to approach them or jump into the social fray, however, and may well have appeared like the stereotypical scientific nerd to them.

Our first destination was a biological station called La Selva on the east coast of Costa Rica, which required first a long bus ride to a river and then a trip up the river in primitive canoes carved out of single tree trunks, although powered by outboard motors. This was high adventure with a profusion of plants and animals unlike anything that I or most of the other students had seen before. For those who love nature as I do, which included most of my new companions, the tropics are infinitely more exciting than a man-made attraction such as Disney World can ever hope to be. Within minutes of heading up the river the entire class began to shout with the unselfconscious wonder of children. "Look at that!" "Oh, wow!" "Incredible!" The teachers smiled, since we were only repeating the exclamations of previous classes and their own first exposure to tropical life. Giant iguanas looking like dragons basked in the trees, butterflies with wings like stained-glass windows fluttered along the banks, gaudy parrots hurtled through the air like bullets with wings, howler monkeys bellowed from the canopy, and the trees were so tall and so cloaked with vines and other plants upon their branches that every square inch seemed to be occupied.

The next morning began with a lecture after breakfast on a topic that could be illustrated with the local fauna and flora. Then we split into groups that actually conducted experiments, gathering and analyzing data to be reported and discussed when the groups reconvened at the end of the day. We were not here just to watch but to test the latest scientific hypotheses about how organisms evolve and interact with each other in their natural environment. The range of questions was breathtaking. One group would place different kinds of bait on the forest floor to see which species of ants were attracted, while another group would travel to a place where a giant tree had fallen to see which species of plants were the first to respond to the sudden increase in light. A third group would open vials containing fragrant chemicals and watch brilliant metallically colored bees appear, while a fourth group would study mites that live in flowers and disperse on the bills of hummingbirds. The next day brought more questions, more experiments, more amazing organisms. This was science in fast motion and it was almost as intoxicating to me and my newfound companions as the tropics themselves.

For the first time, I was with a group of people who were like myself. After the day's work we continued to talk into the night, drinking beer after beer. The conversation wandered from science to other topics as if there were no boundary between the two realms. We were especially interested in the lives we could expect for ourselves as scientists. I was not the only one who had experienced the loneliness of pursuing what those around me could not understand. To my amazement, I discovered that my weaknesses had turned into strengths. During one late-night discussion someone asked who would continue in science if they knew it would make them unhappy. No thought was required for me to raise my hand, but somehow this raised their esteem for me rather than lowering it.

The tropics, the science, and the newfound companionship produced a feeling of intimacy that was new for all of us. On our first trip to the ocean we found ourselves on a sandy beach that stretched as far as the eye could see in both directions without another person in sight. The sky was spotlessly blue, the wind caressed our bodies, and the surf beckoned us to enter. Two of the women exchanged glances, laughed, took off their clothes, and sauntered into the water as if it was the most natural act in the world. Soon everyone was following their example, laughing at this perfect expression of the freedom they were feeling in so many other respects. It was liberating rather than sexual, but for me the sight of so many naked women frolicking in the surf was as if I had truly died and gone to heaven. I took off everything but my glasses, absurdly pretending I was so blind that I couldn't even go swimming without them!

As we traveled from site to site, one woman named Anne increasingly attracted my attention. She was not among the first I had noticed, but she was fun and interesting to talk to, and the more I watched the more I liked her. She had a way of seeing what needed to be done and helping without a hint of pretense. She fit my image of Anne of Green Gables, a farm girl with her hair in a braid and a face upon which makeup would be inappropriate. To my surprise, the more I liked her the more physically appealing she became, while the women who initially grabbed my attention became uglier. Anne was not merely nice; she was beautiful. That experience was the inspiration for my research on beauty that I described in Chapter 16.

Other men in the class were also attracted to Anne and occupied center stage with an ease that I could only envy as I watched from the wings. In despair I imagined her comparing me to dusk as she strode off with night or day, but one afternoon she came into my room where I was resting and sat on the edge of my cot. Our talk deepened, she started to stroke my chest with her fingers, and I pulled her toward me with the words "Oh, Anne!"

Words cannot describe the remainder of the course, when the pieces of my life fit together like a puzzle that until then had defied solution. Glorious nature, glorious science, glorious friends, and glorious Anne! A typical day might find me in the morning by a little stream in an arid part of Costa Rica, studying ant lions. These fascinating insects look like dragonflies in their adult form, but their larvae dig pits in the sand to capture ants and other insects that fall in. They dig the pit by using their oversized jaws like a shovel, flicking the sand over their heads. They lie concealed at the bottom of the pit and when an ant tries to scramble out, the ant lion flicks more sand to create miniature avalanches that carry their victims back into their waiting jaws. After sucking out the juices of its prey, the ant lion delivers a final flick, tossing the carcass out of the pit. Merely by collecting the carcasses that accumulated around the rims, I could measure the feeding success of each ant lion.

As the day grew hot I would take a break by walking down the stream until it passed though a tunnel made of giant slablike rock formations leaning against each other. Bats lining the ceiling fluttered their wings as I passed through the tunnel to a lovely pool on the other side. Removing my clothes, I slipped into the crystal water and swam to the outflow of the pool, which was the top of a waterfall so high that hawks were circling below me. It was the dry season and only a trickle of water passed over the falls, so I could safely lie on my stomach at the very edge while tiny fish fearlessly nibbled the hairs on my leg. This was truly heaven on earth, and the incredible vista below seemed like the vista that had opened upon my life. In the evening there would be mounds of beans and rice to satisfy my enormous hunger, beer and tropical fruit juices to quench my thirst, great boisterous conversation, and the nightly search for a private spot where Anne and I could embrace and talk until sleep could be put off no longer.

When I asked Anne during one of our nightly conversations what attracted her to me, she replied that it was my passion for science, which enabled me to understand her own. Also, she liked the fact that I did not take center stage in social interactions. Even though she could play the social game and, as a woman, was often placed on center stage whether she liked it or not, she usually preferred to be alone or with a few good friends. More of my weaknesses had miraculously turned into strengths. This was what she liked, being with one person and talking about what matters.

Anne and I did not talk about staying together after the course. I was in the middle of a divorce and she was in the middle of her own difficult relationship back home. Instead, in my eternal foolishness I thought that I had discovered an elixir of love that would result in an endless succession of Annes. Alas, whatever worked in Costa Rica didn't work in East Lansing, but I had discovered an elixir of science that continued to work. I was no longer a zooplankton ecologist. I was an everything ecologist, interested in the entire diversity of life. I had discovered the power of evolutionary thinking that I have now passed on to you in this book.

In Costa Rica I had noticed something about ant lions that was similar to the zooplankton I had studied at Woods Hole. In both cases, small food items could be eaten by both large and small individuals while large food items could be eaten only by large individuals. This asymmetrical relationship between prey size and predator size was quite general, as I discovered when I read studies of other predators and their prey. Something similar even applied to human tools; a large pair of forceps can pick up a small item almost as well as a small pair of forceps, but a small pair of forceps is incapable of picking up a large item.

As simple as this observation might seem, it had been ignored by evolutionists who were trying to explain how different species coexist with each other. The prevailing theory was that species could coexist by becoming different in size, with each species having access to resources unavailable to the others. The most important insights are often also the simplest, and it seemed that my simple observation might have important implications for theories of how species coexist.

These theories were expressed in mathematical form. Math had defeated me in college, but I was slowly coming to realize that ideas can be expressed much more precisely in mathematical form than verbally. Evolutionary theory had gone mathematical long ago, and I had to speak the lingo to join the club. Out came my old textbook and new books with reassuring titles such as *Calculus for Idiots*. I kept a huge pad of paper the size of a newspaper on my desk and filled it with equations representing my simple observation that big animals eat big and little while little animals eat only little. The equations looked impressive to those who didn't understand them, but in reality my skills never went beyond what I should have learned in my freshman calculus course. No matter; it was the *idea* that counted, not the math. Give a bad idea to a good mathematician, and he roars off in his mathematical sports car to the middle of nowhere. I might be driving a jalopy, but at least I knew where I was going.

My mathematical model showed that species could coexist by becoming different in size only under special circumstances. I submitted it to a prestigious scientific journal called the *American Naturalist* and it was accepted. From a mediocre student who was rejected by all but my "safety" colleges, I was becoming quite respectable. I was publishing papers in the best journals while still a graduate student, which was a very good sign. I was beginning to speak the lingo of mathematics, which makes one appear like an oracle to those who can't. And the power of adaptationism allowed me to roam the length and breadth of the plant and animal kingdom like a giant with massive strides.

I was not alone. Some of my fellow graduate students at Michigan State were also on fire with evolutionary thinking and publishing in the best journals. Costa Rica had fanned my fire, but theirs had been fanned in other ways. Our little group was not prestigious, consisting of only a few young professors and their students, but we were brash and not intimidated by even the greatest authorities of our day. Most evolutionists are wonderfully egalitarian and even the most eminent would cheerfully come to East Lansing to talk about their work for little more than their travel expenses. I and my fellow graduate students would often set upon them like a team of young lawyers, challenging them on every point. They in turn would be pleased that their work had generated so much interest, and when we left the biology building it was merely to continue the conversation over dinner and beer at the house of one of the professors. Work merged with play, and the people with whom I worked were the same as those with whom I enjoyed relaxing.

Once, much later in life, I sat next to a man on a plane who was very different from me in appearance. He was impeccably dressed in a business suit and wore the kind of conspicuously expensive watch and jewelry that is completely foreign to my tastes. I was dressed in a flannel shirt and a pair of pants bearing the stains of outdoor life. I usually avoid conversations on planes, but he felt like talking and I learned that he was employed by a large corporation with many branch offices. His job was to travel among the branch offices to evaluate their efficiency. Each evaluation would last about six months, during which he would stay in an apartment maintained by the corporation. As the conversation unfolded I began to realize that his business world was devoid of any kind of warm human relationship. His suit and jewelry communicated status and nothing else. I began to realize my own good fortune at having a job that I loved and that enabled me to interact with people on friendly terms. Such simple and natural pleasures, which my companion could not have for any amount of money! When it came time for me to talk about my job, I was surprised at the longing that crept over his face.

My work on body size was intended to be my Ph.D. thesis, but then something else intervened. A paper on vertical migration in zooplankton appeared in the *American Naturalist* that made me think once again about the evolution of altruism. I started to toy with some equations that were similar to the desert island thought experiments described in Chapter 5. At that time, nearly all evolutionists had concluded that between-group selection (the second version of the thought experiment) was invariably weak compared to within-group selection (the first version), so that selfishness always prevailed in the end. My equations were child's play mathematically, but they seemed to indicate otherwise. Moreover, they appeared to be very general, applying to many species and not just zooplankton. In a fever, I forgot all about body size and wrote up my results as a short paper. Knowing no fear, I called the great Ed Wilson at Harvard University and asked if I could meet with him. I made the long drive from Michigan and was there at the appointed time. Ed's laboratory occupies an entire floor of Harvard's Museum of Comparative Zoology, where he maintains colonies of the many ant species that he studies. A busy man, he graciously gave this audacious graduate student a guided tour, placed me in front of a blackboard, sat down in his chair, and said, "You have twenty minutes." I talked like an auctioneer, and he agreed to consider my article for publication in the *Proceedings of the National Academy of Sciences*, after showing it to some of his colleagues. It became my fourth publication.

My new interest created a problem for my Ph.D. research. I no longer cared about body size and had only written a single slim paper on group selection, my new love. My major professor was an aquatic ecologist named Don Hall, a delightful free spirit who had given me free rein up to this point. He and the committee overseeing my work accepted my slim article, making it probably the shortest Ph.D. thesis in the history of evolutionary biology (eleven pages).

One reason that I became so passionate about group selection was because it so clearly related to the human condition, in addition to the rest of life. My professors and peers regarded themselves as *evolutionary biologists*. They respected the academic convention that studying humans is somehow not biology, as if we were set apart from the rest of nature. I had become an *evolutionist*, perhaps because I am the son of a novelist. For me it was an unexpected homecoming. I had been paddling away from my father, but now I had returned to ponder our own species, just like him, except through the lens of evolutionary theory rather than the lens of fictional narrative.

Only one more thing remained to make my life complete. I finally got smart enough to realize that a relationship is an interaction between two people and that Anne was the perfect person with whom to spend the rest of my life.

Thus ends my story about how I became a scientist, to the best of my recollection. What can it tell us in general about the difference between science and non-science, professional and amateur, smart and dumb? I don't know if Viktor E. Frankl ever used the *E*-word, but he's right that the search for meaning is a fundamental human desire that cuts across each of those categories. I can add that evolutionary theory enhances the search for meaning and I hope you agree on the basis of this book. It enhanced Darwin's search from the moment he had his "aha" experience in the carriage, Alfred Russel Wallace's search from the moment he had his aha experience in the Malayan rain forest, and the search of everyone else who has had their aha experience ever since. Anyone can have this experience, regardless of intelligence or level of education.

Moreover, becoming a scientist doesn't require great intelligence, as strange as that might sound. I am not quick at math, statistics, or any of the other trappings of science. I learned them when there was a good reason to, which is the only way we learn anything. The main thing I had in abundance that enabled me to become a scientist was the desire to become a scientist.

Using evolutionary theory to aid one's search for meaning should be regarded as like creative writing. It provides its own reward, even if you can't get paid for it. I'm serious when I say that no matter what you do for a living—even if you are a public relations man with your foot on a used condom—you can become a highly literate evolutionist and even contribute to the professional literature. That is how Alfred Russel Wallace and Beatrix Potter indulged their interest in natural history, and how most novels and short stories are written today by authors who aren't in it for the money.

36

Bon Voyage

FEBRUARY 3, 2006. I was relaxing with Anne and a historian named Daniel Smail at home at the end of a long day. I first learned about Dan when a publisher asked me to review his proposal for a book titled *Outlines for a Deep History of Humankind*. Yes, it has come to the point where a former zooplankton ecologist is consulted to evaluate a book on human history. In his proposal, Dan pointed out that many history texts still act as if history began around 4000 BC in the Middle East. The date bears a disturbing resemblance to Bishop James Usher's early-seventeenth-century estimate of 4004 BC as the year of the earth's origin. The location bears a disturbing resemblance to the reputed location of the Garden of Eden. Most historians are not young earth creationists, of course, but even after the discovery of deep time in the nineteenth century, they still found ways to conceptualize human history as something that began at a certain time and place—the origin of civilization, the origin of writing, the earliest records of the exploits of great men. Dan argued that none of these reasons made any sense. The study of human history should begin in Africa and requires a solid understanding of proximate biological mechanisms such as neurophysiology in addition to what he called the "new ecologies of the Postlithic era."

Huzzah! When I checked into Dan's background, I discovered that he was highly regarded for his work on medieval European history and had just been lured away from Fordham University by Harvard. His interest in what he called deep history had grown largely out of his teaching activities. As I checked the bibliography of his book proposal, I saw that he had done an admirable job of reading the literature that was new for him and familiar to me, but there was still much to discuss. I contacted him by e-mail and invited him to visit Binghamton as an EvoS seminar speaker, an offer that he enthusiastically accepted.

EvoS is our campuswide evolutionary studies program, which I described in previous chapters as a new island of the Ivory Archipelago. We have organized ourselves so that students and faculty alike can learn the basic principles of evolution and embark upon their own voyages of discovery, just as Dan was doing on his own. The hub of the program is the EvoS seminar series, which brings speakers such as Dan to campus at roughly two-week intervals. The speakers are like travelers from other islands of the Ivory Archipelago, coming to pay us a visit. Dan, a historian, was preceded by a legal scholar and would be followed by a drosophila geneticist. The reason that we can host people from so many islands is because everyone is speaking a common language, the language of evolutionary theory. It is really quite remarkable, when you think about it. A university as a single intellectual community. Undergraduate students, graduate students, and professors from all departments getting together to listen with pleasure and understanding to scientists and scholars representing all subject areas. That is the ideal of a liberal arts education, but it cannot be realized in the absence of a common language. We aim to make the Ivory Archipelago the *United* Ivory Archipelago.

Dan actually visited Binghamton University several years ago as a guest of our highly regarded Center for Medieval and Renaissance Studies (CEMERS). He had a good time talking with his specialist colleagues, but his seminar was attended by only about a dozen people. That's typical of modern academic life, but Binghamton has become different, thanks to

EvoS. Dan still had a good time talking with his specialist colleagues, but his seminar filled a large lecture hall. It was followed by a boisterous reception with pizza and beer, which in turn was followed by a continuing discussion attended by over sixty graduate and undergraduate students, who surrounded Dan in a big semicircle. They are taking a course designed around the seminar series that has become hugely popular, and there was nothing required about the spontaneous applause that ended the discussion.

While we were relaxing at home, Dan told me something disturbing about history departments all around the country. Even though the study of history needs to become *deeper*, in fact it is becoming more *shallow* for purely practical reasons. Faced with shrinking budgets, early history is being neglected in favor of recent history, which appears more relevant. In many departments, history might as well have begun only a few hundred years ago. This is tragically shortsighted from an evolutionary perspective. Human history is the fossil record of cultural evolution, as we have seen. We need armies of scholars studying every culture as far back as possible to understand how societies function and change. If universities aren't up to the task, then others must rise to the challenge or the information will be lost forever. That's why there is an important role for amateurs, if they can be appropriately trained and organized, as I emphasized in the previous chapter. If you love history and culture, then you could contribute to such an effort.

February 9, 2006. I was in the office of the superintendent of the Binghamton City School District. No, I didn't do anything bad. I was there to discuss some Web-based software that we have been developing at the university that enables us to visualize the city of Binghamton in a new way. You see a map on the screen, just like any other map. You can view the whole city or zoom in to inspect any particular part in more detail, like a microscope with different degrees of magnification. As you zoom closer, the familiar street names appear. There is a menu bar that enables you to display the locations of schools, parks, fire stations, and the like. What's new and wonderful about the map is that it enables you to display information about the city that would be difficult or impossible to visualize otherwise. For example, if you click "average age" on the menu bar, the city instantly becomes a mosaic of patches that range from white to black, indicating the average age of people living in a given region. Click "average income" and a new mosaic appears. This information comes from the United States Census Bureau and is available for all cities and towns, but the software enables us to actually *see* patterns of variation, as opposed to being overwhelmed by pages and pages of numbers. I love showing this software to people who know the city well. They invariably become excited and want to explore the meaning of the mosaic patterns, just as the early colonists of this region loved mapping the rivers and valleys.

My main interest in visiting the superintendent was to interest her in gathering new information about the social welfare of our city. I told her about my analysis of Mike's beeper study and how prosociality can either thrive or wither, depending upon the local social environment. If we could get the same kind of information for the city of Binghamton that Mike and Barbara gathered for their nationwide study, we could actually *see* the spatial distribution of prosociality and its underlying factors on a block-by-block level. If we repeat the measurements at three-year intervals, we can examine how these patterns change over time. We can attempt interventions in some neighborhoods and actually assess whether they make a difference, compared to similar neighborhoods that did not receive the intervention.

There is no need to mention the *E*-word, not that the superintendent would mind. Facts are common property that can be used by any theory, as I stressed in Chapter 4. I was here to talk about gathering, visualizing, and analyzing facts that might help us enhance the welfare of our neighborhoods, and nobody understood the need for that better than the school superintendent.

April 3, 2006. Good news! My newest project on religion has been funded by the John Templeton Foundation. This one will examine religious conceptions of the afterlife from a cultural evolutionary perspective. Instead of just me and my band of undergraduate students, I will collaborate with Bill Green, whose work on religious conceptions of altruism was described in Chapter 30, to assemble a team of world-class evolutionists interested in religion and world-class religious scholars interested in evolution. We will develop a protocol for studying conceptions of the afterlife in terms of the five major evolutionary hypotheses and the distinction between proximate and ultimate causation that I outlined in Chapters 29 and 30. The same protocol can be used to study many other elements of religion. The study of religion from an evolutionary perspective is on its way to becoming a recognized field of scientific inquiry.

April 11, 2006. I was walking to our tree house down the sloping field of our property, along a stone wall built by long-departed farmers. When I entered the deep shade of the hemlock forest, it was as if people never existed. I climbed the ladder and started a fire in the woodstove. As the warmth spread, I took off my coat and turned on my laptop to put the finishing touches on this book. I gazed out the window at the stream shimmering in the sunlight. I couldn't see most of them, but the patch of stream and forest taken in by my gaze included literally thousands of species, from Dicty the cellular slime mold to burying beetles in their underground chambers and wood turtles emerging from their winter sleep. Each is a living sculpture, endowed by natural selection with the ability to survive and reproduce in unbroken lineages that stretch back to the origin of life.

It's awesome and humbling to contemplate that we are the product of that same sculpting action, not only our bodies but also our minds and the very thoughts that run through our minds. I sometimes wonder what it must have been like to be present during the early days of Darwin's theory, when the idea was so new and so much remained to be discovered. Then I realize that I *am* present during the early days of Darwin's theory. The intellectual events taking place right now are as foundational as the events of 150 years ago. How amazing that virtually everyone can partake in the excitement, as an observer or a participant, as I hope you have seen on the basis of this book. Evolutionary theory is not the kind of belief system that hurls you like an arrow in a previously chosen direction. It is more like a sailboat or a kayak bobbing by the shore, inviting you to take your own voyage of discovery.

ACKNOWLEDGMENTS

Science is a cooperative activity and my debt to others is reflected on every page of this book. Thanks to my colleagues who allowed me to share their work and stories of how they became scientists. Thanks also to the participants of EvoS, including the students in my courses who read chapters while writing was in progress. The entire manuscript was read by Larry Flammer, Rebecca Moldover, Elliott Sober, Lisa Strick, Betty Wilson, Geoff Wilson, and Tamar Wilson, who span the range from homemaker to high school teacher to philosopher. Michelle Tessler, Bill Massey, and Beth Rashbaum, my agent and editors, respectively, made me feel as comfortable publishing a trade book as I have with my previous academic books.

Most authors thank their spouses, but few get to write about them as much as I have written about my wife, Anne Clark. Our personal and intellectual lives are thoroughly entwined and I wouldn't want it any other way. Thanks also to our two daughters, Tamar and Katie, those scrupulous promise keepers who shared our world and are now forging worlds of their own.

NOTES

I have tried to make this book as friendly as possible to readers who have no previous background in science or evolution. My goal is to awaken interest in these subjects and convince readers from all backgrounds that they can understand, enjoy, and even contribute to scientific knowledge from an evolutionary perspective.

Academic books and scientific journal articles are often difficult to read because they delve into the details. The details can be boring for those who do not understand or care about their relevance, but they become fascinating for those who do. Scientists and scholars even use the phrase "the beauty is in the details" to describe the pleasure of going beyond superficial knowledge.

The endnotes and bibliography serve as a portal to the scientific and scholarly literature on the subjects covered in this book. In addition to consulting these references, you should do some exploring on your own. Most scientists maintain informative Web sites, including articles that can be downloaded free of charge, which can be found by typing their name into a search engine such as Google. A specialized branch of Google called Google Scholar (http://scholar. google.com) provides easy access to the scientific and scholarly literature. If you are associated with an institution of higher education, you probably have electronic access to hundreds of journals whose articles can be searched and downloaded in the comfort of your own home or office. Of course, a trip to the library is a pleasure in its own right. Enjoy!

Chapter 1: The Future Can Differ from the Past

The article titled "The New Creationism: Biology Under Attack" is by Barbara Ehrenreich and Janet McIntosh (1997; the quoted passage is on p. 12). Political scientist Ian Lustick's bleak diagnosis of the social sciences is in Lustick (2005; the quoted passage is on p. 1). Steve Pinker provides a book-length account in *The Blank Slate: The Modern Denial of Human Nature*. Rebecca Stott's *Darwin and the Barnacle* (2004) provides a charming description of Darwin's empire of thought. A more detailed description of EvoS, Binghamton University's campus-wide evolutionary studies program, and my "Evolution for Everyone" course is provided in Wilson (2005).

Chapter 2: Clearing the Deck

Robert Richards's book *Darwin and the Emergence of Evolutionary Theories of Mind and Behavior* (1987) provides an excellent scholarly account of how Darwin's ideas were interpreted by Herbert Spencer and others. Scientific theories of gender differences are described in Barbara Ehrenreich and Deirdre English's *For Her Own Good: Two Centuries of the Experts' Advice to Women* (2005). "The eye theory" is described in William Kelleher's *The Troubles in Ballybogoin* (2003).

Chapter 3: A Third Way of Thinking

Two good basic tutorials on evolution are *The Blind Watchmaker* by Richard Dawkins (1996) and *What Evolution Is* (2002) by Ernst Mayr. At a more advanced level, see philosopher Elliott Sober's *The Nature of Selection* (1984). Edited books on infanticide in animals include Van Schaik and Janson (2000) and Hausfater and Hrdy (1984). The quote from Darwin comes from his autobiography (Darwin 1987/1958, p. 20).

Chapter 4: Prove It!

Jean-Henri Fabre was a great writer in addition to a great naturalist. *Fabre's Book of Insects* (1998) is still available in paperback. Anyone with a serious interest in creationism should read Ronald Number's *The Creationists: The Evolution of Scientific Creationism* (1992).

Chapter 5: Be Careful What You Wish For

Sarah Blaffer Hrdy (1979) encountered stiff opposition to her ideas, and her review article on infanticide is still worth reading. T. H. Huxley, "Darwin's Bulldog," was among those who thought that evolution leaves little room for optimism. The Princeton University Press edition of Huxley's essay "Evolution and Ethics" includes a commentary by the eminent contemporary evolutionist George C. Williams, who feels the same way. As George puts it, "Mother Nature is a wicked old witch." My three desert island thought experiments introduce the concept of *multilevel selection*, which will be elaborated throughout the book. The first half of *Unto Others*, my book with Elliott Sober, discusses multilevel selection theory in detail, including its turbulent history. A more recent article titled "Human Groups as Adaptive Units: Toward a Permanent Consensus" can be downloaded from my Web site. Visit William Muir's Web site at Purdue University to learn more about those good hens!

Chapter 6: Monkey Madness

An excellent lecture by Elliott Sober on the inability of creationism and intelligent design theory to predict the properties of organisms can be viewed on the EvoS Web site. A more scientifically detailed account of the mad monkey story is provided by Suomi (2005).

Chapter 7: How the Dog Got Its Curly Tail

The silver fox experiment is entertainingly described in an *American Scientist* article by Trut (1999; the passage quoted is on p. 170). The best way to familiarize yourself with Doug Emlen's work is to visit his Web site. The idea that natural selection is often constrained by development was a favorite theme of Stephen Jay Gould in his many books of essays. A good recent popular book on evolution and development is Sean B. Carroll's *Endless Forms Most Beautiful: The New Science of Evo Devo and the Making of the Animal Kingdom* (2005). Darwin's observation about drooping ears is in the first chapter of *Origin of Species* (p. 101 of the edition edited by J. Carroll [Darwin 1859/2003]).

Chapter 8: Dancing with Ghosts

The idea that organisms are often adapted to their past but not their present environment is called "mismatch theory" and is frequently invoked by evolutionary psychologists. An article titled "Ecological and Evolutionary Traps" (Schlaepfer, Runge, and Sherman 2002) reviews examples for both humans and other species. John Byer's popular science book *Built for Speed: A Year in the Life of Pronghorn* (2003) is an entertaining portrait of scientists at work studying a species that is "overbuilt" for its current environment. The adaptive behavioral flexibility exhibited by the wood frog and minnow is called phenotypic plasticity and is an extremely important principle of evolution; see Massimo Pigliucci's *Phenotypic Plasticity: Beyond Nature and Nurture* (2001). The phenotypic plasticity of the developing fetus and its health consequences is described in detail in *The Fetal Matrix: Evolution, Development, and Disease* (2004) by Peter Gluckman and Mark Hanson.

Chapter 9: What Is the Function of a Can Opener? How Do You Know?

Evolutionists Stephen Jay Gould and Richard Lewontin wrote a famous essay in 1979 titled "The Spandrels of San Marco and the Panglossian Paradigm: A Critique of the Adaptationist Program" that criticized some of their colleagues for relying too heavily on natural selection thinking and not paying enough attention to constraining factors such as development and phylogeny. I have tried to provide a balanced perspective in this book, first emphasizing the importance of natural selection thinking (Chapter 4) and then emphasizing the constraining influences (Chapters 6 to 8).

Chapter 10: Your Apprentice License

Adaptationism and Optimization (2001), edited by Steve Orzack and Elliott Sober, provides an advanced discussion of how the various elements of evolutionary theory can be combined to form an effective research program.

Chapter 11: Welcome Home, Prodigal Son

My chapter in *The Literary Animal* (2005) is titled "Evolutionary Social Constructivism" and expands upon these themes. It is available on my Web site. My evolutionist colleague Henri Plotkin (1994) uses the term "Darwin machines" to describe evolutionary processes built by evolution, including the immune system, psychological processes, and cultural processes. Gerald Edelman explains his theory of neural Darwinism to a general audience in his recent book *Wider than the Sky* (2005).

Chapter 12: Teaching the Experts

The profile of Margie Profet was in the June 1996 issue of *Scientific American*. Her scientific review of pregnancy sickness was published in an edited book titled *The Adapted Mind* (Barkow, Cosmides, and Tooby 1992), which helped to establish the field of evolutionary psychology. Flaxman and Sherman (2000) on pregnancy sickness and Billing and Sherman (1998) on spices are the first scientific papers that I assign to my students, in part because they provide such good examples of hypothesis formation and testing. To learn how other aspects of medicine and health are being approached from an evolutionary perspective, try *Why We Get Sick* by Randy Nesse and George Williams (1995) or *Evolution in Health and Disease*, edited by Stephen Stearns (1999).

Chapter 13: Murder City

The Human Behavior and Evolution Society (http://www.hbes.com) is an excellent way to become more involved in evolutionary studies. The Web site is informative, and amateurs are welcome to join in addition to professional academics from all disciplines. Membership includes a subscription to the journal *Evolution and Human Behavior* and the annual meetings are a great social and intellectual experience. The passage quoted on page 87 is from Gelles and Straus (1985, p. 88). Daly and Wilson's work on infanticide has recently been critiqued by David Buller in his book *Adapting Minds: Evolutionary Psychology and the Persistent Quest for Human Nature* (2005). The title of Buller's book is a reference to the influential edited volume *The Adapted Mind* (Barkow, Cosmides, and Tooby 1992), which helped to establish the field of evolutionary psychology. The seemingly slight difference between the titles reflects foundational differences in how the human mind is envisioned from an evolutionary perspective. *The Adapted Mind* portrays the mind as a large number of specialized "modules" that evolved by genetic evolution to solve specific problems of survival and reproduction. Open-ended psychological and cultural processes are largely ignored, although acknowledged as a possibility. *Adapting Minds* emphasizes the more open-ended processes. Both of these conceptions have an element of truth, and I have tried to provide examples of both in this book. Buller rejects some of Daly and Wilson's specific claims about infanticide, but his own interpretation is also fully evolutionary. These differences of opinion among evolutionists are a good example of science in progress and show how much remains to be discovered in the future. I have tried to emphasize the provisional state of our current knowledge throughout my book.

Chapter 14: How I Learned to Stop Worrying and Love Genetic Determinism

My chapter in *The Literary Animal* is titled "Evolutionary Social Constructivism" and expands upon these themes. It is available on my Web site. Philosopher Paul Boghossian's book *Fear of Knowledge: Against Relativism and Constructivism* (2006) provides a concise and very effective critique of the anything-goes version of social constructivism. The passage by Margo and Martin is from page 1271 of their article (Wilson and Daly 1997).

Chapter 15: They've Got Personality!

A little article that we wrote titled "Shyness and Boldness in Humans and Other Animals" (Wilson et al. 1994) shows how new these ideas were only a decade ago. Look up Niels Dingemanse on the Internet for a more detailed description of current research on tits, including downloadable articles. More on extroversion and introversion in humans can be found in Nettle (2005), Aron and Aron (1997), and Aron (1996). The passage by Victor Frankl is from his book *Man's Search for Meaning* (2000, pp. 55–6). A branch of mathematics called game theory is often used to model the evolution of social interactions, leading to the coexistence of more than one behavioral strategy in the population (Maynard Smith 1982; Axelrod 1984; Dugatkin 1999; Gintis 2000).

Chapter 16: The Beauty of Abraham Lincoln

Books that explore aesthetics from an evolutionary perspective include *Biophilia* (Wilson 1984), *The Biophilia Hypothesis* (Kellert and Wilson 1993), and *Evolutionary Aesthetics* (Voland and Grammer 2003). Our study on the importance of nonphysical factors in the perception of physical attractiveness (Kniffin and Wilson 2004) is available on my Web site.

Chapter 17: Love Thy Neighbor Microbe

The quote from Evans-Pritchard is from page 64 of *Theories of Primitive Religion* (1965). Gregory Velicer (2003) provides an excellent review of altruism and selfishness in microbes in his article titled "Social Strife in the Microbial World." Paul Rainey's experiments on the mat-forming bacteria are described in Rainey and Rainey (2003). Joan Strassman and David Queller (2004) study cooperation and conflict in Dicty the cellular slime mold. All of them maintain informative Web sites with downloadable articles that you can access by typing their names into a search engine such as Google.

Chapter 18: Groups All the Way Down

The story about the elderly woman is the scientific equivalent of an urban legend. The famous scientist varies among versions, from William James to Thomas Huxley to Bertrand Russell. Lynn Margulis explains her ideas in *The Origin of Eukaryotic Cells* (1970) and *Symbiotic Planet: A New Look at Evolution* (1998), among other books. Her quote about human consciousness and spirochete ecology is from Barlow (1991, p. 55). John Maynard Smith and Eörs Szathmáry developed the concept of major transitions in *The Major Transitions of Life* (1995) and *The Origins of Life: From the Birth of Life to the Origins of Language* (1999). A more technical discussion of the origin of life is provided in Szathmáry et al. (2005).

Chapter 19: Divided We Fall

Austin Burt and Robert Trivers provide a comprehensive review of genetic evolution and conflict within individual organisms in their book *Genes in Conflict: The Biology of Selfish Genetic Elements* (2006). My account of cancer as a process of evolution among cell lineages is based on an article by Bernard Crespi and Kyle Summers (2005) titled "The Evolutionary Biology of Cancer."

Chapter 20: Winged Minds

In addition to his book *The Wisdom of the Hive* (2001), Tom Seeley has written a charming essay titled "A Feeling, and a Fondness, for the Bees" that describes his approach to studying a single species intensively as a model organism. Visit his Web site to download this and other scientific research articles that show how "the beauty is in the details." Another good book on social insect colonies as self-organized systems is *Self-Organization in Biological Systems* (2001) by Scott Camazine et al.

Chapter 21: The Egalitarian Ape

The account of the !Kung San tribesman is from pages 45 to 47 of Boehm (1999). The passages about Inuits and Ona are on pages 68 and 62 respectively. Other accounts of hunter-gatherer life include Lee and Devore (1976), Turnbull (1987), and Howell (1984). In addition to his earlier books *Chimpanzee Politics* (1982) and *Good Natured* (1995), Frans de Waal's most recent books include *Bonobo: The Forgotten Ape* (1998), *My Family Album: Thirty Years of Primate Photography* (2003), and *Our Inner Ape: A Leading Primatologist Explains Why We Are Who We Are* (2005). Although Chris Boehm relies heavily on the comparison between humans and chimpanzees, we are equally closely related to the bonobos, who use sex as a social currency, including sex among females to form alliances that resist domination by males. In general, the social systems of our closest relatives are extraordinarily diverse. Paul Bingham's stone-throwing theory is outlined in Bingham (1999).

Chapter 22: Across the Cooperation Divide

Michael Tomasello and his colleagues at the Max Planck Institute for Evolutionary Anthropology are prolific researchers, publishing several papers a year on the topics explored in this chapter. Their most recent article as of this writing is titled "What's in It for Me? Self-regard Precludes Altruism and Spite in Chimpanzees" (Jensen et al. 2006). Mike provides a broad overview of his work in *Behavioral and Brain Sciences* (Tomasello et al. 2005). This is the journal that I described and analyzed in Chapter 1, which consists of target articles followed by commentaries, providing a comprehensive exploration of a given subject area. Other articles that provided material for this chapter include Hare et al. (2002), Kobayashi and Kohshima (2001), and Miklosi et al. (2003).

Chapter 23: The First Laugh

In addition to Matt's article (Gervais and Wilson 2005), I recommend *Laughter: A Scientific Investigation* (2000) by Robert Provine, who has been studying laughter from a scientific and evolutionary perspective for many years. The concept of mirror neurons as a "shared manifold of intersubjectivity" is developed by Gallese (2003). The importance of positive emotions for individual and social development is discussed by Frederickson (1998).

Chapter 24: The Vital Arts

The tiny but fascinating literature on subjects associated with the humanities from an evolutionary perspective includes Dissanayake (1990, 1995, 2000) and Miller (2000) for art in general, McNeill (1995) for dance, Brown (2000) and Wallin et al. (2001) for music, Coe (2003) for visual art, and Carroll (1994, 2004) and Gottschall and Wilson (2005) for literature. The passages from McNeill (1995) are on pages 2, 8, 9, and 10 respectively. The passages from Brown (2000) are on pages 237 and 238. The passages from Coe (2003) are on pages 25 to 27. The passage from Gottschall and Wilson is on page xviii.

Chapter 25: Dr. Doolittle Was Right

In addition to Terry Deacon's *The Symbolic Species* (1998), Sue Savage-Rumbaugh and her collaborators describe their research with Kanzi and other apes in *Apes, Language, and the Human Mind* (2002). You can read more about Alex the talking—and thinking— parrot in Irene Pepperberg's *The Alex Studies: Cognitive and Communicative Abilities of African Grey Parrots* (2002).

Chapter 26: How Many Inventors Does It Take to Make a Lightbulb?

The word "groupthink" was coined by Irving L. Janis in *Groupthink: Psychological Studies of Policy Decisions and Fiascoes* (1992). My two articles (Wilson 1997; Wilson, Timmel, and Near 2004) provide extensive references to the scientific literature on group cognition. Two psychologists who appreciate the advantages of thinking in groups are Daniel Wegner (1986) and Edwin Hutchins (1995). Other articles that I drew upon for this chapter are Kruglanski and Webster (1991), Tetlock et al. (1992), Michaelson et al. (1989, 1992), and Ziman (2003).

Chapter 27: I Don't Know How It Works!

Many people associate cultural evolution with the "memes," a term coined by Richard Dawkins in his book *The Selfish Gene* (1976). Unfortunately, the term does not have a precise meaning. Used broadly, it becomes a synonym for "any cultural trait," much as the term "selfish gene" becomes a synonym for "any gene that evolves," as I mentioned in Chapter 6. Used more narrowly, it implies that cultural traits can be decomposed into discrete genelike units and also that they can evolve to be organisms in their own right that don't necessarily benefit their human "hosts." These ideas are developed by authors such as Susan Blackmore (1999) and Robert Aunger (2002). In my opinion, Peter Richerson and Robert Boyd provide the best overview of the modern study of cultural evolution in *Not by Genes Alone: How Culture Transformed Human Evolution* (2005). The passage from Wilson and Holldobler (2005) is on page 13371. The passages from Ong (1998) are on pages 33, 35, 51, 54, and 56. The passage from Achebe (1961/1994) is on page 57. Passages from Nisbett and Cohen (1996) are on pages 2, 87, 86. The idea that cultural diversity in America can be attributed largely to cultural diversity among the colonists is documented by historian David Hackett Fischer in *Albion's Seed: Four British Folkways in America* (1989).

Chapter 28: Darwin's Cathedral

Recent books on religion from an evolutionary perspective include Scott Atran's *In Gods We Trust: The Evolutionary Landscape of Religion* (2002), Pascal Boyer's *Religion Explained* (2001), Daniel Dennett's *Breaking the Spell: Religion as a Natural Phenomenon* (2006), Robert Hinde's *Why Gods Persist: A Scientific Approach to Religion* (1999), and Richard Dawkins's *The God Delusion* (2006). Well-known religious scholars such as Elaine Pagels (1995, 2003), Jack Miles (1996), and Rodney Stark and William Bainbridge (1985, 1987) don't frame their arguments in terms of evolution but these arguments can clearly be given an evolutionary formulation. Game theory is an approach to modeling social behavior that involves pitting various social strategies against each other. Successful strategies such as "tit-for-tat" combine traits such as niceness, retaliation, and forgiveness in a way that bears a striking resemblance to the Hutterite passage that I quoted at length. See Axelrod (1980), Dugatkin (1999), Gintis (2000), and Maynard Smith (1982) for more on game theory. The passages quoted from *Darwin's Cathedral* are on pages 89, 123, 214 to 215, and 183. The Hutterite passages are from Ehrenpreis (1650/1978), pages 11 and 66 to 69. The passage from Darwin (1887/1958) is on page 70.

Chapter 29: Is There Anyone Out There? Is There Anyone Up There?

The five-CD Ray Charles collection is *Ray Charles: Genius and Soul, the 50th Anniversary Collection*, from Rhino Records. Elliott Sober, the philosopher with whom I wrote *Unto Others*, objects to my terms "factual realism" and "practical realism" because the term "realism" usually implies factual realism in the philosophical literature. Nevertheless, I think that the terms are appropriate because practical beliefs must be anchored to factual reality in some sense, even if they do not directly represent factual reality. I highly recommend *The Universe in a Single Atom: The Convergence of Science and Spirituality* (2005), by His Holiness the Dalai Lama, which shows how science and religion can be brought harmoniously together. Myles Horton's *The Long Haul: An Autobiography* (1990) shows how one person found a purely practical belief in "Love thy neighbor" to be fully motivating. The passage quoted is from page 7. The passage from Benjamin Franklin is from his autobiography and quoted on page 25 of Norman Cousins's book *In God We Trust: The Religious Beliefs and Ideas of the American Founding Fathers* (1958), which shows why the framers of the Constitution were so insistent about the separation of church and state. The definition of Islam is from Eliade (1987, volume 7, page

119). Passages from the Dalai Lama (2005) are on pages 76, 59, 52, 144, and 148.

Chapter 30: Ayn Rand: Religious Zealot

Scientific theories that support conventional beliefs about the role of women in society are described by Barbara Ehrenreich and Deirdre English in *For Her Own Good: Two Centuries of the Experts' Advice to Women* (2005), as I mentioned in the endnote to Chapter 2. I develop the concept of multiple "species of thought" in two articles (Wilson 1990, 1995). John Allman's book *Evolving Brains* (1999) shows how much the brain of an electric eel differs from that of a bat (for example), illustrating my statement that "deception begins with perception." The article titled "How Love Came Back" was published in the September 1986 issue of *Reader's Digest*. The passage from the Alcoholics Anonymous manual is on page 62. The passages from Seabury (1937/1964) are on pages 11, 15 to 16, 27, and viii. The passage from Rand (1961) is on page 50. Passages from Branden (1989) are on pages 7, 15, and 18. The passage by Alan Greenspan is from the April 24, 1974, issue of *Newsweek*.

Chapter 31: The Social Intelligence of Nations, or, Evil Aliens Need Not Apply

The fields of economics and political science are only beginning to be approached from an evolutionary perspective, as political scientist Ian Lustick describes in the passage quoted in Chapter 1. Some key books include *Evolutionary Foundations of Economics* (Dopfer 2005), *Evolutionary Psychology and Economic Theory* (Koppl 2005), *Economics and Evolution* (Hodgson 1995), *Microeconomics: Behavior, Institutions, and Evolution* (Bowles 2003), *Moral Sentiments and Material Interests: The Foundations of Cooperation in Economic Life* (Gintis et al. 2005), *Darwinian Politics: The Evolutionary Origin of Freedom* (Rubin 2002), *Darwinian Conservatism* (Arnhart 2005), and *A Darwinian Left: Politics, Evolution, and Cooperation* (Singer 2000). Specialized journals include *Politics and the Life Sciences* and *Bioeconomics*. Authors such as Arnhart and Singer attempt to show how positions associated with current-day American conservatism and liberalism can be given an evolutionary formulation. I think that evolutionary theory offers a more even playing field, as I have attempted to show in my chapter. The passage from Mandeville (1705/1957) is on page 19. The passage from Smith (1759/1976) is on pages 184 to 185.

Chapter 32: Mr. Beeper

Among Mike Csikszentmihalyi's many books, I recommend *The Evolving Self: A Psychology for the Third Millennium* (1993), which shares some of the themes in this book. Robert Wright's book *Nonzero: The Logic of Human Destiny* (2002) attempts to give Teilhard de Chardin's thesis a modern evolutionary formulation. It is similar in some respects to Howard Bloom's *Global Brain: The Evolution of the Mass Mind from the Big Bang to the 21st Century* (2000). All three authors think that the human race is on its way to becoming one big planetary organism. I think that this is naive. Groups *can* evolve to be like organisms, as I show in this book, but special conditions are required, especially at the scale of the entire planet. In addition to technological advances in communication such as the Internet, we need to create a true global village with the balance of power that naturally exists in a small-scale human society, as I argue in the previous chapter. In the absence of these conditions, cultural evolution will lead in the direction of inequality, such as the concentration of wealth that I describe in this chapter. I criticize Wright's book in more detail in a review titled "Nonzero and Nonsense: Group Selection, Nonzerosumness, and the Human Gaia Hypothesis" (Wilson 2000). Sir John Templeton's books include *Discovering the Laws of Life* (1994) and *Worldwide Laws of Life* (1997) Passages from Teilhard de Chardin (1955/1976) are on pages 224 and 251 to 252.

Chapter 33: The Ecology of Good and Evil

My first article on this research project is titled "Health and the Ecology of Altruism" (Wilson 2006) and only begins to scratch the surface of this wonderful database. We are currently looking at the effects of religious experience in more detail, and we have also conducted our own study with four weeks of beeper data, providing over two hundred snapshots of experience for each individual.

Chapter 34: Mosquitoes Under the Bed

I explore the ethical implications of evolutionary theory in more detail in two articles titled "Evolution, Morality, and Human Potential" (Wilson 2002) and "On the Inappropriate Use of the Naturalistic Fallacy in Evolutionary Psychology" (Wilson et al. 2003). The passage from McCullough (1978) is on page 145. The widely quoted remark "Nothing in biology makes sense except in the light of evolution" is from Dobzhansky (1973). Passages from Lemov (2005) are on pages 21, 41, 74, 57 to 58, 208, 222, and 234.

Chapter 35: The Return of the Amateur Scientist

A recent biography of Alfred Russel Wallace is *In Darwin's Shadow: The Life and Science of Alfred Russel Wallace* (2002) by Michael Shermer. An article in *Natural History* magazine titled "The Other World of Beatrix Potter" (Whalley 1988) describes her talents as a mycologist. I highly recommend the North Country School (http://www.nct.org/school) and its affiliated summer camp, Camp Treetops, for a wonderful learning and living experience. The Organization for Tropical Studies (http://www.ots.duke.edu/) is still going strong and is very worthy of your support and involvement.

Chapter 36: Bon Voyage

Dan Smail's book *Outlines for a Deep History of Humankind* will be published by the University of California Press. Part of his argument is presented in an article titled "In the Grip of Sacred History," published in the *American Historical Review* (Smail 2005).

BIBLIOGRAPHY

Achebe, C. (1958/1994). *Things Fall Apart*. New York, Anchor.

———. (1961/1994). *No Longer at Ease*. New York, Anchor.

Almann, J. M. (1999). *Evolving Brains*. New York, Scientific American Library.

Arnhart, L. (2005). *Darwinian Conservatism*. Exeter, UK, Imprint Academic.

Aron, E. N. (1996). *The Highly Sensitive Person: How to Thrive When the World Overwhelms You*. New York, Broadway.

Aron, E. N., and A. Aron (1997). "Sensory-Processing Sensitivity and Its Relation to Introversion and Emotionality." *Journal of Personality and Social Psychology* 73:345–68.

Atran, S. (2002). *In Gods We Trust: The Evolutionary Landscape of Religion*. Oxford, UK, Oxford University Press.

Aunger, R. (2002). *The Electric Meme*. New York, Free Press.

Axelrod, R. (1984). *The Evolution of Cooperation*. New York, Basic Books.

Barkow, J. H., L. Cosmides, and J. Tooby, eds. (1992). *The Adapted Mind: Evolutionary Psychology and the Generation of Culture*. Oxford, UK, Oxford University Press.

Barlow, C. (1955). *From Gaia to Selfish Genes: Selected Writings in the Life Sciences*. Cambridge, Mass., MIT Press.

Berreby, D. (2005). *Us and Them: Understanding Your Tribal Mind*. New York, Little, Brown.

Billing, J., and P. W. Sherman (1998). "Antimicrobial Function of

Spices: Why Some Like It Hot." *Quarterly Review of Biology* 73: 3–49.

Bingham, P. M. (1999). "Human Uniqueness: A General Theory." *Quarterly Review of Biology* 74:133–69.

Blackmore, S. (1999). *The Meme Machine*. Oxford, UK, Oxford University Press.

Bloom, H. (2000). *Global Brain: The Evolution of Mass Mind from the Big Bang to the 21st Century*. New York, Wiley.

Boehm, C. (1984). *Blood Revenge*. Philadelphia, Penn., University of Pennsylvania Press.

———. (1999). *Hierarchy in the Forest*. Cambridge, Mass., Harvard University Press.

Bowles, S. (2003). *Microeconomics: Behavior, Institutions, and Evolution*. Princeton, NJ, Princeton University Press.

Boyer, P. (2001). *Religion Explained*. New York, Basic Books.

Branden, N. (1989). *Judgement Day*. Boston, Mass., Houghton Mifflin.

Brown, S. (2000). "Evolutionary Models of Music: From Sexual Selection to Group Selection." *Perspectives in Ethology* 13:231–81.

Buford, B. (1990). *Among the Thugs*. London, UK, Trafalgar Square.

Buller, D. J. (2005). *Adapting Minds: Evolutionary Psychology and the Persistent Quest for Human Nature*. Cambridge, Mass., MIT Press.

Burt, A., and R. Trivers (2006). *Genes in Conflict: The Biology of Selfish Genetic Elements*. Cambridge, Mass., Belknap Press.

Byers, J. A. (2003). *Built for Speed: A Year in the Life of Pronghorn*. Cambridge, Mass., Harvard University Press.

Camazine, S., J.-L. Deneubourg, N. R. Franks, J. Sneyd, G. Theraulaz, and E. Bonabeau (2001). *Self-organization in Biological Systems*. Princeton, N.J., Princeton University Press.

Cannon, W. B. (1939). *Wisdom of the Body*. New York, W. W. Norton.

Carroll, J. (1994). *Evolution and Literary Theory*. Columbia, Mo., University of Missouri Press.

———. (2004). *Literary Darwinism: Evolution, Human Nature, and Literature*. Oxford, UK, Routledge.

Carroll, S. B. (2005). *Endless Forms Most Beautiful: The New Science of Evo Devo and the Making of the Animal Kingdom*. New York, W. W. Norton.

Coe, K. (2003). *The Ancestress Hypothesis: Visual Art as Adaptation.* New Brunswick, NJ, Rutgers University Press.

Cousins, N. (1958). *In God We Trust: The Religious Beliefs and Ideas of the American Founding Fathers.* New York, Harper.

Crespi, B., and K. Summers (2005). "Evolutionary Biology of Cancer." *Trends in Ecology and Evolution* 20:545–52.

Csikszentmihalyi, M. (1990). *Flow: The Psychology of Optimal Experience.* New York, Harper and Row.

———. (1993). *The Evolving Self: A Psychology for the Third Millennium.* New York, HarperCollins.

Csikszentmihalyi, M., and B. Schneider (2000). *Becoming Adult: How Teenagers Prepare for the World of Work.* New York, Basic Books.

Dalai Lama (2005). *The Universe in a Single Atom.* New York, Morgan Road Books.

Daly, M., and M. Wilson (1988). *Homicide.* New York, Aldine de Gruyter.

Darwin, C. (1859/2003). *On the Origins of Species* (J. Carroll, ed.). Toronto, Canada, Broadview Press.

———. (1871). *The Descent of Man and Selection in Relation to Sex.* New York, Appleton.

———. (1887/1958). *The Autobiography of Charles Darwin, 1809–1882. With Original Omissions Restored.* New York, Harcourt Brace.

Dawkins, R. (1976). *The Selfish Gene.* Oxford, UK, Oxford University Press.

———. (1996). *The Blind Watchmaker: Why the Evidence of Evolution Reveals a Universe Without Design.* New York, W. W. Norton.

———. (2006). *The God Delusion.* New York, Houghton Mifflin.

Deacon, T. W. (1998). *The Symbolic Species.* New York, W. W. Norton.

Dennett, D. C. (2006). *Breaking the Spell: Religion as a Natural Phenomenon.* New York, Viking.

Diamond, J. (1997). *Guns, Germs, and Steel.* New York, W. W. Norton.

Dissanayake, E. (1990). *What Is Art For?* Seattle, Wash., University of Washington Press.

———. (1995). *Homo Aestheticus: Where Art Comes from and Why.* Seattle, Wash., University of Washington Press.

———. (2000). *Art and Intimacy: How the Arts Began.* Seattle, Wash., University of Washington Press.

Dobzhansky, T. (1973). "Nothing in Biology Makes Sense Except in the Light of Evolution." *American Biology Teacher* 35:125 to 9.

Dopfer, K., ed. (2005). *The Evolutionary Foundations of Economics.* Cambridge, UK, Cambridge University Press.

Dugatkin, L. A. (1999). *Cheating Monkeys and Citizen Bees.* New York, Free Press.

Durkheim, E. (1912/1995). *The Elementary Forms of Religious Life.* New York, Free Press.

Edelman, G. (2005). *Wider than the Sky: The Phenomenal Gift of Consciousness.* New Haven, Conn., Yale University Press.

Ehrenpreis, A. (1650/1978). "An Epistle on Brotherly Community as the Highest Command of Love." In *Brotherly Community: The Highest Command of Love.* Rifton, NY, Plough.

Ehrenreich, B., and D. English (2005). *For Her Own Good: Two Centuries of the Experts' Advice to Women.* New York, Anchor.

Ehrenreich, B., and J. McIntosh (1997). "The New Creationism: Biology Under Attack." *The Nation,* June 9:11–16.

Eliade, M., ed. (1987). *The Encylopedia of Religion.* New York, Macmillan.

Evans-Pritchard, E. E. (1965). *Theories of Primitive Religion.* Oxford, UK, Clarendon Press.

Fabre, J.-H. (1998). *Fabre's Book of Insects.* New York, Dover.

Fischer, D. H. (1989). *Albion's Seed: Four British Folkways in America.* New York, Oxford University Press.

Fisher, R. A. (1930/1958). *The Genetical Theory of Natural Selection.* New York, Dover.

Flaxman, S. M., and P. W. Sherman (2000). "Morning Sickness: A Mechanism for Protecting Mother and Embryo." *Quarterly Review of Biology* 75:113–48.

Frankl, V. (2000). *Man's Search for Meaning.* Boston, Beacon Press.

Frederickson, B. L. (1998). "What Good Are Positive Emotions?" *Review of General Psychology* 2:300–19.

Fukuyama, F. (1992). *The End of History and the Last Man.* New York, Free Press.

————. (1995). *Trust*. New York, Free Press.

Gallese, V. (2003). "The Roots of Empathy: The Shared Manifold Hypothesis and the Neural Basis of Intersubjectivity." *Psychopathology* 36:171–80.

Gelles, R. J., and M. A. Strauss (1985). "Violence in the American Family." In *Crime and the Family*, (A. J. Lincoln and M. A. Strauss, eds.). Springfield, Ill., Thomas, 88–110.

Gervais, M., and D. S. Wilson (2005). "The Evolution and Functions of Laughter and Humor: A Synthetic Approach." *Quarterly Review of Biology* 80:395–430.

Gintis, H. (2000). *Game Theory Evolving*. Princeton, N.J., Princeton University Press.

Gintis, H., S. Bowles, R. Boyd, and E. Fehr, eds. (2005). *Moral Sentiments and Material Interests: The Foundations of Cooperation in Economic Life*. Cambridge, Mass., MIT Press.

Gluckman, P., and M. Hanson (2004). *The Fetal Matrix: Evolution, Development, and Disease*. Cambridge, UK, Cambridge University Press.

Gottschall, J., and D. S. Wilson, eds. (2005). *The Literary Animal: Evolution and the Nature of Narrative*. Evanston, Ill., Northwestern University Press.

Gould, S. J., and R. C. Lewontin (1979). "The Spandrels of San Marco and the Panglossian Paradigm: A Critique of the Adaptationist Program." *Proceedings of the Royal Society of London* B205:581–98.

Gray, J. G. (1998). *The Warriors: Reflections on Men in Battle*. Lincoln, Neb., University of Nebraska Press.

Hare, B., M. Brown, C. Williamson, and M. Tomasello (2002). "The Domestication of Social Cognition in Dogs." Science 298:1634–6.

Hausfater, G., and S. B. Hrdy, eds. (1984). *Infanticide: Comparative and Evolutionary Perspectives*. New York, Aldine.

Hinde, R. (1999). *Why Gods Persist: A Scientific Approach to Religion*. New York, Routledge.

Hodgson, G. M. (1993). *Economics and Evolution*. Cambridge, UK, Polity Press.

Horton, M. (1990). *The Long Haul: An Autobiography*. New York, Doubleday.

Howell, S. (1984). *Society and Cosmos: Chewong of Peninsular Malaya*. Singapore, Oxford University Press.

Hrdy, S. B. (1979). "Infanticide Among Animals: A Review, Classification, and Examination of the Implications for the Reproductive Strategies of Females." *Ethology and Sociobiology* 1:13–40.

Hutchins, E. (1995). *Cognition in the Wild*. Cambridge, Mass., MIT Press.

Huxley, T. H. (1989). *Evolution and Ethics: T. H. Huxley's Evolution and Ethics with New Essays on Its Victorian and Sociobiological Context*. Princeton, N.J., Princeton University Press.

Janis, I. L. (1982). *Groupthink: Psychological Studies of Policy Decisions and Fiascoes*. New York, Houghton Mifflin.

Jenson, K., B. Hare, J. Call, and M. Tomasello (2006). "What's in It for Me? Self-regard Precludes Altruism and Spite in Chimpanzees." *Proceedings of the Royal Society of London* 273:1013–21.

Kelleher, W. F. J. (2003). *The Troubles in Ballybogoin*. Ann Arbor, University of Michigan Press.

Kellert, S. R., and E. O. Wilson, eds. (1993). *The Biophilia Hypothesis*. Washington, D.C., Shearwater.

Kelly, R. C. (1985). *The Nuer Conquest*. Ann Arbor, Mich., University of Michigan Press.

Kniffin, K., and D. S. Wilson (2004). "The Effect of Non-physical Traits on the Perception of Physical Attractiveness: Three Naturalistic Studies." *Evolution and Human Behavior* 25:88–101.

Kobayashi, H., and D. Kohshima (2001). "Unique Morphology of the Human Eye and Its Adaptive Meaning: Comparative Studies on External Morphology of the Primate Eye." *Journal of Human Evolution* 40:419–435.

Koppl, R., ed. (2005). *Evolutionary Psychology and Economic Theory*. Greenwich, Conn., JAI Press.

Kruglanski, A. W., and D. M. Webster (1991). "Group Members' Reactions to Opinion Deviates and Conformists at Varying Degrees of Proximity to Decision Deadline and of Environmental Noise." *Journal of Personality and Social Psychology* 61:212–25.

Lansing, J. S. (1991). *Priests and Programmers: Technologies of Power in the Engineered Landscape of Bali*. Princeton, N.J., Princeton University Press.

Lee, R. B., and I. DeVore (1976). *Kalahari Hunter-Gatherers: Studies of the !Kung San and Their Neighbors*. Cambridge, Mass., Harvard University Press.

Lemov, R. (2005). *World as Laboratory: Experiments with Mice, Mazes, and Men*. New York, Hill and Wang.

Lewis, S. (1925). *Arrowsmith*. New York, Harcourt Brace.

Lustick, I. S. (2005). "Daniel Dennett, Comparative Politics, and the Dangerous Idea of Evolution." *APSA-CP Newsletter* 16(2):1–8.

Mandeville, B. (1705/1957). *The Fable of the Bees: or Private Vices, Public Benefits*. Oxford, UK, Clarendon Press.

Margulis, L. (1970). *Origin of Eukaryotic Cells*. New Haven, Conn., Yale University Press.

———. (1998). *Symbiotic Planet: A New Look at Evolution*. New York, Basic Books.

Maynard Smith, J. (1982). *Evolution and the Theory of Games*. Cambridge, UK, Cambridge University Press.

Maynard Smith, J., and E. Szathmáry (1995). *The Major Transitions of Life*. New York, W. H. Freeman.

———. (1999). *The Origins of Life: From the Birth of Life to the Origin of Language*. Oxford, UK, Oxford University Press.

Mayr, E. (2002). *What Evolution Is*. New York, Basic Books.

McCullough, D. (1978). *Path Between the Seas*. New York, Simon and Schuster.

McGrath, A. E. (1990). *A Life of John Calvin*. Oxford, UK, Basil Blackwell.

McNeill, W. H. (1995). *Keeping Together in Time: Dance and Drill in Human History*. Cambridge, Mass., Harvard University Press.

Michaelsen, L. K., W. E. Watson, and R. H. Black (1989). "A Realistic Test of Individual Versus Group Consensus Decision Making." *Journal of Applied Psychology* 74:834–39.

Michaelsen, L. K., W. E. Watson, A. Schwartzkopf, and R. H. Black (1992). *Journal of Applied Psychology*. "Group Decision Making:

How You Frame the Question Determines What You Find."
77:106–08.

Miklosi, A., E. Kubinyi, J. Topal, M. Gacsi, Z. Viranyi, and V. Csanyi (2003). "A Simple Reason for a Big Difference: Wolves Do Not Look Back at Humans, but Dogs Do." *Current Biology* 13:763–6.

Miles, J. (1996). *God: A Biography*. New York, Vintage.

Miller, G. (2000). *The Mating Mind: How Sexual Choice Shaped the Evolution of Human Nature*. New York, Doubleday.

Morris, D. (1967). *The Naked Ape: A Zoologist's Study of the Human Animal*. New York, McGraw-Hill.

Nesse, R. M., and G. C. Williams (1995). *Why We Get Sick: The New Science of Darwinian Medicine*. New York, Crown.

Nettle, D. (2005). "An Evolutionary Approach to the Extroversion Continuum." *Evolution and Human Behavior* 26:363–73.

Neusner, J., and W. C. Green (2005). *Altruism in World Religions*. Washington, D.C., Georgetown University Press.

Nisbett, R. (2003). *Geography of Thought: How Asians and Westerners Think Differently, and Why*. New York, Free Press.

Nisbett, R. E., and D. Cohen (1996). *Culture of Honor*. New York, Westview Press.

Numbers, R. L. (1992). *The Creationists: The Evolution of Scientific Creationism*. Berkeley, Calif., University of California Press.

Ong, W. J. (1998). *Orality and Literacy: The Technologizing of the Word*. Oxford, UK, Routledge.

Orzack, S. H., and E. Sober, eds. (2001). *Adaptationism and Optimality*. Cambridge, UK, Cambridge University Press.

Pagels, E. (1995). *The Origin of Satan*. Princeton, N.J., Princeton University Press.

———. (2003). *Beyond Belief: The Secret Gospel of Thomas*. New York, Random House.

Pepperberg, I. M. (2002). *The Alex Studies: Cognitive and Communicative Abilities of Grey Parrots*. Cambridge, Mass., Harvard University Press.

Pigliucci, M. (2001). *Phenotypic Plasticity: Beyond Nature and Nurture*. Baltimore, Md., John Hopkins University Press.

Pinker, S. (2002). *The Blank Slate: The Modern Denial of Human Nature*. Viking, New York.

Plotkin, H. (1994). *Darwin Machines and the Nature of Knowledge.* Cambridge, Mass., Harvard University Press.

Profet, M. (1992). "Pregnancy Sickness as an Adaptation: A Deterrent to Maternal Ingestion of Teratogens." In *The Adapted Mind: Evolutionary Psychology and the Generation of Culture* (J. H. Barkow, L. Cosmides, and J. Tooby, eds.). Oxford, UK, Oxford University Press, 327–65.

———. (1995). *Protecting Your Baby to Be: Preventing Birth Defects in the First Trimester.* New York, Perseus Books.

Provine, R. R. (2000). *Laughter: A Scientific Investigation.* London, Penguin.

Putnam, R. D. (1992). *Making Democracy Work: Civic Traditions in Modern Italy.* Princeton, N.J., Princeton University Press.

Rainey, P. B., and K. Rainey (2003). "Evolution of Cooperation and Conflict in Experimental Microbial Populations." *Nature* 425:72–74.

Rand, A. (1947). *The Fountainhead.* New York, Bobbs-Merrill.

———. (1957). *Atlas Shrugged.* New York, Random House.

———. (1961). *The Virtue of Selfishness.* New York, Signet.

Reagan, M., and S. Begley (2002). *Inside the Mind of God: Images and Words of Inner Space.* West Conshohocken, Penn., Templeton Foundation Press.

Richerson, P. J., and R. Boyd (2004). *Not by Genes Alone: How Culture Transformed Human Evolution.* Chicago, Ill., University of Chicago Press.

Rubin, P. (2002). *Darwinian Politics: The Evolutionary Origin of Freedom.* New Brunswick, N.J., Rutgers University Press.

Sapolsky, R. M. (1998). *Why Zebras Don't Get Ulcers.* New York, W. H. Freeman.

Savage-Rumbaugh, S., S. G. Shanker, and T. J. Taylor (2001). *Apes, Language, and the Human Mind.* Oxford, UK, Oxford University Press.

Seabury, D. (1937/1964). *The Art of Selfishness.* New York, Pocket Books.

Schlaepfer, M. A., M. C. Runge, and P. W. Sherman (2002). "Ecological and Evolutionary Traps." *Trends in Ecology and Evolution* 17: 474–480

Seeley, T. (1995). *The Wisdom of the Hive*. Cambridge, Mass., Harvard University Press.

———. (2001). "A Feeling, and a Fondness, for the Bees." In *Model Systems in Behavioral Ecology* (L. A. Dugatkin, ed.). Princeton, N.J., Princeton University Press.

Shermer, M. (2002). *In Darwin's Shadow: The Life and Science of Alfred Russel Wallace*. Oxford, UK, Oxford University Press.

Singer, I. B. (1962). *The Slave*. New York, Farrar, Straus and Giroux.

Singer, P. (2000). *A Darwinian Left: Politics, Evolution, and Cooperation*. New Haven, Conn., Yale University Press.

Smail, D. (2005). "In the Grip of Sacred History." *American Historical Review* 110:1336–61.

Smail, D. (in press). *Outlines for a Deep History of Humankind*. Berkeley, Calif., University of California Press.

Smith, A. (1759/1976). *Theory of Moral Sentiments*. Oxford, UK, Oxford University Press.

Sober, E. (1984). *The Nature of Selection*. Cambridge, Mass., MIT Press.

Sober, E., and D. S. Wilson (1998). *Unto Others: The Evolution and Psychology of Unselfish Behavior*. Cambridge, Mass., Harvard University Press.

Stark, R. and W. S. Bainbridge (1985). *The Future of Religion*. Berkeley, Calif., University of California Press.

———. (1987). *A Theory of Religion*. New Brunswick, N.J., Rutgers University Press.

Stearns, S. C., ed. (1999). *Evolution in Health and Disease*. Oxford, UK, Oxford University Press.

Stott, R. (2004). *Darwin and the Barnacle: The Story of One Tiny Creature and History's Most Spectacular Scientific Breakthrough*. New York, W. W. Norton.

Strassman, J. E., and D. C. Queller (2004). "Sociobiology Goes Micro: Long Used for Studying Development, Dictyostelium Now Also Provides a Model for Analyzing Social Interactions." *ASM News* 70:526–32.

Suomi, S. J. (2005). "Genetic and Environmental Factors Influencing the Expression of Impulsive Aggression and Serotonergic

Functioning in Rhesus Monkeys." In *Developmental Origins of Aggression* (R. E. Tremblay, W. H. Hartup, and J. Archer, eds.). New York, Guilford Press, 63–82.

Szathmáry, E., M. Santos, and C. Fernando (2005). "Evolutionary Potential and Requirements for Minimal Protocells." *Topics in Current Chemistry* 259:169–211.

Teilhard de Chardin, P. (1955/1976). *The Phenomenon of Man*. New York, Harper.

Templeton, J. M. (1994). *Discovering the Laws of Life*. West Conshohocken, Penn., Templeton Foundation Press.

———. (1997). *Worldwide Laws of Life*. West Conshohocken, Penn., Templeton Foundation Press.

Tetlock, P. E., R. S. Peterson, C. McGuire, S. Chang, and P. Feld (1992). "Assessing Political Group Dynamics: A Test of the Groupthink Model." *Journal of Personality and Social Psychology* 63:403–25.

Tocqueville, A. de (1835/1990). *Democracy in America*. Garden City, N.Y., Anchor.

Tomasello, M., M. Carpenter, J. Call, T. Behne and H. Moll (2005). "Understanding and Sharing Intentions: The Origins of Cultural Cognition." *Behavioral and Brain Sciences* 28:675–735.

Trut, L. N. (1999). "Early Canid Domestication: The Farm-Fox Experiment." *American Scientist* 87:160–70.

Turnbull, C. M. (1987). *The Forest People*. Carmichael, Calif., Touchstone.

Van Schaik, C. P., and C. H. Janson, eds. (2000). *Infanticide by Males and Its Implications*. Cambridge, UK, Cambridge University Press.

Velicer, G. J. (2003). "Social Strife in the Microbial World." *Trends in Microbiology* 11:330–37.

Voland, E., and K. Grammer, eds. (2003). *Evolutionary Aesthetics*. Berlin, Springer-Verlag.

Vonnegut, K. (1963). *Cat's Cradle*. New York, Laurel.

de Waal, F. (1992). *Chimpanzee Politics*. Baltimore, Md., John Hopkins University Press.

———. (1995). *Good Natured*. Cambridge, Mass., Harvard University Press.

———. (2003). *My Family Album: Thirty Years of Primate Photography*. Berkeley, Calif., University of California Press.

———. (2005). *Our Inner Ape: A Leading Primatologist Explains Why We Are Who We Are*. New York, Riverhead.

de Waal, F., and F. Lanting (1998). *Bonobo: The Forgotten Ape*. Berkeley, Calif., University of California Press.

Wallin, N. L., B. Merker, and S. Brown, eds. (2001). *The Origins of Music*. Cambridge, Mass., MIT Press.

Wegner, D. M. (1986). "Transactive Memory: A Contemporary Analysis of the Group Mind." In *Theories of Group Behavior* (B. Mullen and G. R. Goethals, eds.). New York, Springer-Verlag.

Weiner, J. (1994). *The Beak of the Finch: A Story of Evolution in Our Time*. New York, Knopf.

Wesley, J. (1976). "Thoughts upon Methodism." In *The Works of John Wesley* (R. E. Davies, ed.). Nashville, Tenn., Abington Press.

Whalley, J. I. (1988). "The Other World of Beatrix Potter." *Natural History* 97:48–52.

Wilson, D. S. (1990). "Species of Thought: A Comment on Evolutionary Epistomology." *Biology and Philosophy* 5:37–62.

———. (1995). "Language as a Community of Interacting Belief Systems: A Case Study Involving Conduct Toward Self and Others." *Biology and Philosophy* 10:77–97.

———. (1997). "Incorporating Group Selection into the Adaptationist Program: A Case Study Involving Human Decision Making." In *Evolutionary Social Psychology* (J. Simpson and D. Kenrick, eds.). Mahwah, N.J., Erlbaum, 345–86.

———. (2000). "Nonzero and Nonsense: Group Selection, Nonzerosumness, and the Human Gaia Hypothesis." *Skeptic* 8:84–89.

———. (2002). *Darwin's Cathedral: Evolution, Religion, and the Nature of Society*. Chicago, Ill., University of Chicago Press.

———. (2002). "Evolution, Morality and Human Potential." In *Evolutionary Psychology: Alternative Approaches* (S. J. Scher and F. Rauscher, eds.,). Boston, Mass., Kluwer, 55–70.

———, ed. (2004). *The New Fable of the Bees*. Advances in Austrian Economics. Greenwich, Conn., JAI Press.

————. (2005). "Evolution for Everyone: How to Increase Acceptance of, Interest in, and Knowledge About Evolution." *Public Library of Science (PLoS) Biology* 3:1001–8.

————. (2006). "Human Groups as Adaptive Units: Toward a Permanent Consensus." In *The Innate Mind: Culture and Cognition* (P. Carruthers, S. Laurence, and S. Stich, eds.). Oxford, U.K., Oxford University Press.

Wilson, D. S., A. B. Clark, K. Coleman, and T. Dearstyne (1994). "Shyness and Boldness in Humans and Other Animals." *Trends in Ecology and Evolution* 9:442–6.

Wilson, D. S., and M. Csikszentmihalyi (2006). "Health and the Ecology of Altruism." In *The Science of Altruism and Health*, (S. G. Post, ed.). Oxford, UK, Oxford University Press.

Wilson, D. S., E. Dietrich, and A. B. Clark (2003). "On the Inappropriate Use of the Naturalistic Fallacy in Evolutionary Psychology." *Biology and Philosophy* 18:669–82.

Wilson, D. S., J. Timmel, and R. R. Miller (2004). "Cognitive Cooperation: When the Going Gets Tough, Think as a Group." *Human Nature* 15:225–50.

Wilson, E. O. (1971). *The Insect Societies.* Cambridge, Mass., Belknap Press.

Wilson, E. O. (1975). *Sociobiology: The New Synthesis.* Cambridge, Mass., Harvard University Press.

————. (1984). *Biophilia.* Cambridge, Mass., Harvard University Press.

————. (1998). *Consilience.* New York, Knopf.

Wilson, E. O., and B. Holldobler (2005). "Eusociality: Origin and Consequences." *Proceedings of the National Academy of Sciences* 102:13367–71.

Wilson, M., and M. Daly (1997). "Life Expectancy, Economic Inequality, Homicide, and Reproductive Timing in Chicago Neighborhoods." *British Medical Journal* 314:1271–8.

Wilson, T. D. (2002). *Strangers to Ourselves: Discovering the Adaptive Unconscious.* Cambridge, Mass., Harvard University Press.

Wright, R. (2000). *Nonzero: The Logic of Human Destiny.* New York, Pantheon.

Ziman, J., ed. (2003). *Technological Innovation as an Evolutionary Process.* Cambridge, UK, Cambridge University Press.

WEB SITES

Most of the scientists discussed in this book maintain informative Web sites describing their research and providing a list of their publications, often as downloadable files. I have not listed the Web sites for individuals because they change fairly often. To find a given individual, simply type his or her name into a search engine such as Google. If there are too many individuals with the same name, add a keyword such as the organism they study. Here is a list of Web sites for organizations that I discuss, which are likely to remain more stable.

Chapter	Web site	Description
1	http://biology.binghamton.edu/dwilson	My Web site, including articles discussed in this book that can be downloaded free of charge
1	http://bingweb.binghamton.edu/~evos/	EvoS, Binghamton University's campuswide evolutionary studies program and a new island of the Ivory Archipelago
3	http://www.scholar.google.com/	Google Scholar, a specialized branch of Google for searching scientific and scholarly literature
13	http://www.hbes.com/	The Human Behavior and Evolution Society
15	http://www.hsperson.com/	Highly Sensitive Person Web site, maintained by Elaine Aron
17	http://dictybase.org/	Dictybase, the only place to be if you want to know about the cellular slime mold

24	http://www.aldaily.com/	Arts and Letters Daily, a Web site maintained by Denis Dutton in collaboration with the *Chronicle of Higher Education*
25	http://www.alexfoundation.org/	The Alex Foundation, where you can support the research of Irene Pepperberg
30	http://www.aynrand.org/	The Ayn Rand Institute
35	http://www.nct.org/school/	The North Country School
35	http://www.ots.duke.edu/	The Organization for Tropical Studies

INDEX